Environmental Geography

Environmental Geography

Mohan Singh

ABD PUBLISHERS
Jaipur ■ New Delhi

ABD PUBLISHERS

Regd. Off. "Bony Residency", Gate No. 2
Opp. Tilak Public School, Vishweswariya Nagar,
Gopalpura Road, Jaipur - 302018, India
Ph: 0141 - 2761280, Telefax: 2761381
Email: abdpublishersjpr@gmail.com

Delhi Office: 102, 1st Floor, Satyam House,
4327/3, Ansari Road, Darya Ganj,
New Delhi - 110002, India
Ph: 011-45652440

visit us: www.oxfordbookcompany.com

Distributed By
KEYA INTERNATIONAL PRESS LLP
E-mail: keyainternationalpressllp@gmail.com

Published in India, 2026

ISBN: 9788183763103

Typeset by Shivangi Computer, Jaipur

Printed and bound in India by Thomson Press Ltd.,
New Delhi

Contents

Preface

Environment is the source of life on the planet, determining the existence, growth and development of mankind and other organisms and its activities. For man, nature is a pool of resource. He has been ignorant and has been exploiting it incessantly without considering the need for conservation until recently but why? Because earlier the rate of restoration was better but gradually human activities began to alter the basic structure and composition of the components and the elements. When the situation became unfavourable, man realized the need for preservation and improvement for his own sake. Environmental changes at any scale, from global to local, are of crucial importance to life on planet Earth. In addition to climate change, the major contemporary challenges include changes in biodiversity, soil degradation, and increases in mass movements. Often the dynamics of such environmental processes require an understanding of the spatial and temporal couplings between the atmosphere, biosphere, relief sphere, pedosphere, and hydrosphere. In addition, the interaction with human activities is of utmost importance. Physical geography – as an interdisciplinary field with a marked focus on the environment-society nexus – is among the core disciplines addressing and examining these environmental changes. The interface of air, water and land forming life supporting layer known as Biosphere is the broadest geoecosystem (ecosystem) which is the spatial unit for the study of environmental geography. The prime concern of environmental geography is thus to study the components of natural environment separately and together their linkage at various levels through environmental and biological processes and human responses to environment vis a vis man environment relationship. Traditionally, geography is mainly considered as the scientific study of various places on Earth. However, the modernistic view not only includes the designations of the natural and artificial landforms on the planet, but also human demography, economic activities, cultural and historical development, social organizations and climates. The interactions between these major elements provide a better understanding on how humans directly or indirectly shape the environment.

Environmental geography focuses on the physical environment and its effect on humans. Understanding the integration of physical and environmental geography has never been more important than it is today to address the environmental challenges facing the world. One of the most important factors in development is geography, where the country is in the world, and climate. It's no coincidence that the poorest countries are in the tropics, where it is hot, the land is less fertile, and water is scarcer, where diseases flourish. Conversely, Europe and North America profit from huge tracts of very fertile land, a temperate climate, and good rainfall. In extremes of climate, either hot or cold, too much energy goes into the simple business of survival for there to be much leftover energy for development. You have to work twice as hard to get enough to eat out of the ground, you have to irrigate where others can depend on rainfall. It may be too hot to work between 11 and 2, so you lose three hours out of every day. Rain patterns may give you a short growing season, while others can get two harvests in one year. Some countries are just at a natural disadvantage. Secondly, geographical location plays a part in access to markets. All the great empires have been based around trade routes, and these are almost always maritime. Many of the world's poorest countries are severely hindered because they are landlocked; situated in high mountain ranges; or lack navigable rivers, long coastlines, or good natural harbours. Thirdly, every country has been dealt a hand in natural resources. It takes infrastructure to capitalise on these, but some places have a distinct advantage over others. Environmental stability can be a factor in development. Some countries are more stable than others.

This valuable book offers comprehensive geographical information on all the aspects of the environment. Graduate students of Geography will find this book helpful in their study and work.

— *Mohan Singh*

Chapter 1

Man and Environmental Relationship

Environment may be defined as the surrounding conditions—physical or non-physical, that influence and modify the life of an individual or a community. It influences the habits and modes of life of man and his degree of civilization. Some groups of people in particular environmental conditions have profound possibilities for economic activities, while others have to work in either limited or adverse conditions. The particular environmental conditions of U.K., Germany and Japan have made them industrially and economically very advanced. The fertile river valley of China and India together with monsoon type of climate have made these countries agriculturally rich. The people of Savannah are, as a rule, the great hunters of the world. The environmental conditions are favourable for wild animals—carnivores and herbivores. The temperate grasslands of the world are the granaries of the world. All the principal wheat lands of the world are associated with these grasslands. The ever hot, ever wet climates have made the Amazonian people indolent and unenterprising. On the other hand, the mild equable climate of Western Europe has produced energetic, active and most enterprising people.

We must not, however, forget that the same type of environment may not necessarily produce a similar mode of living. Some groups of people may derive more benefit from their environment, while others living in similar conditions may not be well advanced in the field of economic activity. The quality and the culture of the people are the dominating factors in the utilization of economic resources in different environmental conditions. The economic development of Australia and South Africa was made not by the local people but by the people of Western Europe, who were endowed with high culture and superior quality. These people are scientifically well advanced. They could make the best utilization of the economic resources available in these regions. The Congo and the Amazon basins, and Indonesia have the same type of environmental conditions. But owing to the fact that the people of Indonesia are more advanced than the other two regions, the country is economically and politically more advanced and developed. Of course, owing to the insular situation of Indonesia, white settlement was possible. The contribution of the white people in developing the region was no less important. So, though the mode of life in any region is influenced by the factors of environment, it is the quality and culture of the people that matter much in carrying on their economic activities.

The factors of environment can broadly be classified into two groups—physical and non-physical.

The principal physical factors are (a) Geographical location, (b) coastline, (c) Topography, (d) inland water bodies, (e) climate, (f) soil, (g) animal, (h) vegetation, (i) minerals, etc.

The principal non-physical factors are (a) Population, (b) Political and Social organization.

Geographical Location

The geographical location of a country in relation to the surface of the globe determines to a great extent, the industrial and commercial development.

The location of a country can be studied from the following points of view:—

The location with reference to land and sea: The location with reference to land and sea may be Insular, Littoral or marginal, Isthmian, Peninsular and Continental.

When a country is surrounded by the sea, the location is called *insular*. This provides access to the sea on all sides of the country. All the ocean routes of the world become accessible to the country. Sea-winds have moderating effect on climate. This type of land produces sea-faring nations. People become hardy and skilful and commercially and industrially advanced. Britain and Japan have insular positions. They have established their trade and commerce through ocean routes with the different countries of the world. Moreover, the defence of a country having a sea-frontier becomes comparatively easier.

Littoral or marginal location means that a considerable portion of a country is located by the side of the sea. The area is influenced by sea winds and can carry on trade along the coast by sea route. France, Norway, Sweden and Spain enjoy this advantage.

An isthmus is a narrow strip of land that connects two larger land masses. The isthmus of Panama is an important example. This connects the two great continents, namely, the North America and the South America. By cutting this isthmus two mighty oceans, namely, the Atlantic and the Pacific have been linked up. This has offered a great possibility of carrying on trade between the western and the eastern coasts of the U.S.A. and the western and northern coastal countries of South America through ocean route.

Peninsular location means that a country or a region is surrounded by the sea on three sides. Southern India, Malyasia, Italy, etc., have this type of location. These countries can carry on trade and commerce both by sea and land-routes. Sea-winds also have influence on their climate. The location of a country is said to be *continental* when it is located far away from the sea surrounded by landmasses. The climate of this region becomes extreme. Summer temperature becomes high and winter temperature becomes low. Sea-borne trade becomes difficult. Contact with foreign countries becomes limited. Defence of land frontier becomes a great problem. Afghanistan, Bolivia, Poland etc., offer good examples of continental location. A country like Nepal has to depend on India to carry on trade with other countries.

Location with reference to latitude: On the variation of latitude, a country may be located in tropical, sub-tropical, temperate or polar regions. Tropical and sub-tropical countries like India, Indonesia, etc., have hot and humid climate. Agricul-turally these countries are rich. The principal agricultural crops are rice, jute, cotton, sugar-cane, rubber, tea, etc. Temperate countries have moderate climate. The people are energetic. Industrially these countries are rich. They produce temperate crops like wheat, barley, sugar-beet, etc. Mechanization in agriculture is important. Sea-fishing near the sea-coasts is an important occupation. Countries like the U.K., Germany, France, Japan, Norway offer good examples. In the polar regions of Canada, Soviet Union, and Greenland economic activity is limited. Seal-fishing and special type of animal hunting are the principal occupations. Eskimos of Canada are important to note.

Location in relation to other Advanced countries: If a country is situated near an industrially and commercially developed country, its economy may be influenced by that country. Western European countries for their industrial and commercial development owe much to the U.K. Progress of newly formed Bangladesh depends much on its relation with India, a more advanced and progressive country, from the point of view of agriculture, industry, trade and commerce. A country remains backward when it fails to come in contact with outside world.

Altitude also a *modifying* effect on climate. Quito is situated on the equator. It should have ever hot, ever wet climate. But because it is situated at a great altitude, its temperature has been modified. Hence it is the land of "*eternal spring.*"

Nature of Coastline

The coastline is a line which indicates the line of separation between land and sea. Therefore, all countries with littoral, peninsular or insular position must have sea-coasts. The coastline has a great contribution in developing ports and harbours. The coastline of a country may be indented (broken) and unindented (straight). The coastline may also be high or low.

The *indented* or *broken* coast has great commercial importance. The broken coast helps intrusion of sea into the land. This becomes ideal for the construction of natural ports and harbours, through which sea-borne trade is carried on. Shipping yard can be developed for accommodating and building ships. Through these ports sea-fishing is developed. The port of Visakhapatnam, Cardiff, etc., are ideal sea-ports on broken coasts. Broken coasts of Greece and Norway have made them a sea-faring nations. Broken coasts bring interior landmass near to the sea. The country with broken coasts has larger numbers of natural ports and harbours. Great Britain has many natural ports and harbours.

The *unindented* or *straight* coast : A coastline without a cut into the land by sea is called unindented coast. Such a coast offers little possibility of developing natural port and harbour. The coastline of Africa is not so indented. India with a coastline of about 6083 km, —because of unindented coast for a long distance—has only a small number of ideal ports and harbours like Bombay, Goa, Cochin, and Visakhapatnam. Madras, an artificial port, entails a great cost for its maintenance.

The coastline may be *high* or *low*. The coastline of Africa is locked by mountains. Access to the sea becomes difficult. Adjacent seas are shallow. Development of ports and harbours is difficult. On the other hand, the coastline of North-Western European countries is low as well as indented, thus offering great facilities for the development of natural port and harbour.

Topography

Topography is the nature of land surface and is a reflection of the Geological structure. Topography of the land surface can be divided into three features; namely, *mountains, plateaus,* and *plains.* These three landforms have great influence on the economic development of a country. Plain lands are more advanced that the mountanious and plateau areas.

Influence of Mountains: Though in general mountains offer a great resistance in developing a region economically, yet many facilities man can derive from the mountains for economic development.

Facilities Offered by the Mountains: (i) Mountains act as natural boundary, between two political units. Mountainous frontier offers impediments to easy access by the enemy. The northern and eastern frontiers of India, locked by the mighty Himalayas and its offshoots are more safe. (ii) Mountains are the sources of great rivers. The mighty rivers like the Ganges, the Brahmaputra and the Indus are fed by Himalayan glaciers and they are perennial. They have profound influence on the economic life of the regions through which they pass. Mountain torrents are highly suitable for the generation of hydroelectricity—the cheapest source of power in modern industry. These hydel powers play the most vital part on the economic life of Japan. India's economy is also highly influenced by the hydel power derived from mountain torrents. (iii) Mountains influence the climate of a country. The South-West monsoon winds striking against the Himalayas supply heavy rain to the different parts of India. On the other hand, the Himalayas protect India from the cold wind of the North. (iv) Mountain at its different heights and slopes contains different types of vegetations from hardwood forest to softwood forest—which have great commercial use. The Himalayas and the Rockies are full of such lucrative forests. (v) Mountain slopes are very ideal for the production of a special variety of agricultural crops such as tea, rubber, coffee, etc. Himalayan slopes are ideal for tea, Malaysian mountain slopes for rubber and Brazilian plateau for coffee. (vi) Various types of economic minerals and rocks are obtained from the mountains. Younger rocks contain oil and coal and harder rocks, contain iron, copper, lead, etc. (vii) High mountain top with beautiful natural scenery and healthy climate offers creation of recreation centres and health resorts. Darjeeling, Simla and Nainital are famous in India. Switzerland with beautiful mountain climate is called the playground of Europe. These health-resorts attract tourists, encourage hotel-keeping. The people of mountainous regions are strong, healthy and brave. The Nepalese are renowned warriors.

Limitations Offered by the Mountains: Mountains in many ways offer great resistance in developing the region. The discontinuous uneven mountain terrain stands in the way of developing communication system. Cost of construction is very high. Movement through mountain terrain is not easy and settlement is difficult. Hence the mountainous region is thinly populated. Landslide, accumulation of huge fost and snow stand in the way of developing the region. Soil erosion on mountain slopes is great. Fertility of soil is easily lost. Agriculture is carried on in restricted areas. In recent times, however, man is trying to use the hill slopes in many ways.

Influence of Plateau: Plateaus have flat surfaces and stand well above the neighbouring lands. The river valleys are deep and steep-sided. The steepness of the plateau edges often makes it difficult to reach the top. From the economic point of view the plateau is superior to mountains but inferior to plains in respect to the possibility for economic development. The plain top of the plateau with greater area can be used for various economic purposes. Development of transport system is easier and except on the plateau slopes, the soil erosion is not so great on the top. Soil is comparatively fertile. Plateau is generally a storehouse of minerals. Western Australian plateau is full of minerals like gold, copper and lead. The swift rivers of the plateau slopes are very ideal for the generation of hydro-electricity. The "fall line" of the Appalachian region of the U.S.A. is an outstanding example of hydro-electric source along the plateau margin. Lava plateau of Western India is a great cotton-growing area.

Influence of Plains: Of the different topographical features, plains play the most vital role in developing the economy of a country. Fifty percent of the world's land surface is occupied by different types of plain and plain is the home of about 90% of the world population. Hence, plain is the most densely populated area with varieties of economic activities.

Plain land offers following facilities for man :—

- The level and more or less even surface of plain with rich fertile soil is ideal for agriculture. Soil erosion is less. The plain is thus ideal for developing irrigation systems with rivers, canals and wells. Some of the great river valleys and deltas of the world, namely, those of the Ganges, the Brahmaputra, the Volga, the Mississippi are great agricultural lands. The plain temperate grasslands of North America and South America are noted for the production of foodgrains and other crops.

- In the plain region development of roads, railways, inland waterways like canals are easier and they together with rivers offer great facilities for transport. The Ganges valley with roads, railways, canals, etc., is well developed from the communication point of view.
- In the plain region industrial, trade and commercial centres, cities and towns are easily built up.

Some desert plains are not suitable for human settlement. But the sandy barren Nile valley of Egypt, Rajasthan area in India have been made fertile by developing irrigation system. In short, the plain region is highly developed from the point of view of agriculture, industry, trade and commerce.

Influence of Island Water Bodies

About 70% of the surface of the earth is occupied by water-bodies. Of these, oceans and rivers are the most important. Inland water-bodies, however, consist of rivers, lakes, canals, ponds, etc. Of these water-bodies, again, rivers and lakes play the most important part in the economic life of man.

Influence of Rivers: Of all the physical factors of environment none has played a more important role than rivers and river valleys in helping man's progress and civilization in many ways. Some of the river valleys of the world, namely, those of the Nile, the Ganges, the Idus, the Tigris and Euphratis, the Hwang Ho, etc., are the cradles of earliest civilization. Most of the river valleys and deltas are thickly populated.

The following benefits are derived from the rivers :

- A river supplies water for drinking and fish for food. The Hooghly is the source of drinking water for the great city of Culcutta. The Ganges, the Brahmaputra, etc. are the sources of fish
- In the upper course, the river is swift and torrential. In this part, the river is destructive in nature. But the swift flow of water is utilized for the generation of water-power, the cheapest source of power in industry today.
- In the middle and lower courses of the river, by constructing irrigation canals water from the river is carried on to the agricultural land for irrigation purpose. This irrigation system

plays the most important part in agriculture. A network of canals in the Punjab, Rajasthan, U.P., etc., carrying water from the rivers have turned the dry and semi-desert areas into principal agricultural land. The Nile with its canals have turned the desert valley and delta into well-developed agricultural land. Hence Egypt is called the gift of the Nile. Had there be no Nile, Egypt would remain a desert land.

- The rivers along with its canals play the most vital part as inland communication system. Cities and towns have been developed on both banks of the river. Some of the great ports of the world such as London, and Calcutta are situated respectively on the Thames and the Hooghly. A river gives passage to the sea—thus helping sea-borne trade.
- A river brings with it huge amount of silt and deposits them on both banks of the river and on deltas making those regions fertile. Hence the river valleys and deltas are agriculturally rich.
- Some of the rivers act as boundary-line between two political units. The Danube of Europe is an international river crossing the boundaries of Germany, Austria, Czechoslovakia, Yugoslavia, Rumania and Bulgaria.
- Some of the rivers offer recreation facilities such as swimming, boating etc.

But some of the rivers by causing flood, erosion etc., bring about miseries to the life and property of the people of the region. The Damodar of West Bengal and Hwang Ho of China were the sorrows of the respective areas through which they pass. But today by developing multipurpose river valley projects some of these rivers have been tamed. By these projects not only has the flood been controlled but also the following objects have been accomplished. These are : (i) Checking of flood and soil erosion, (ii) irrigation facility, (iii) production of water power, (iv) transport facility, etc. In their turn, these have given rise to various other economic activities.

Another source of inland waterways is some of the *great lakes* of the world. In the U.S.A. and Canada, the five Great Lakes play the most important part as inland waterways. Coal, iron ore, nickel, and various types of agricultural and industrial products are carried on by both the countries through these water bodies.

In France, Germany and Soviet Union the network of canal system plays the most important part as inland waterways.

Influence Lakes and Canals

Influence of Lakes

A lake is a sheet of water surround by land. The water occupies a basin in the earth's surface. Lakes may be of different types on the basis of size, shape, depth, amount of salt present with water and mode of formation.

Lakes also play an important part in influencing the economic activites of man.

Influence of Lakes on Climate: Like ocean lakes also as water body influence the climate of a place. Lake water in summer lowers the temperature of air and increases the temperature in winter month. The five great lakes of North America namely, Superior, Michigan, Huron, Erie and Ontario modifies the temperature of the region and by increasing the moisture of air through evaporation help in increasing the rainfall. Caspean Sea is the biggest salt water lake. The Elburz mountain lies South of it. The winds blowing over the Caspean Sea absorb moisture and these moisture ladden winds when strike against the northern slope of the mountain discharge rain water.

Influence of Lakes as Recreation Centre: Lakes play the most important part as recreation centre. Beautiful scenery attracts tourists from different parts. The lakes of Kashmir are very important in this respect.

Sources of Minerals: The water of the lakes are the sources of minerals and fish. By drying the salty water of lakes *salt,* very useful to man, is obtained. Varieties of fish are obtained from the lake. The Chilka lake of India supplies fish which is sent to different states of India.

Uses of Lakes as Inland Water-ways: Lakes, in some parts of the world, play the most vital part as inland water-route. The five great lakes of North America play the most vital part as means of communication. They have been joined by canals to make the water route perfect. Internal and foreign trade of the U.S.A. and Canada is carried on through the route.

Importance of Lakes as Source of Power, Irrigation and Drainage: The great lakes of North America are separated by falls and rapids. The Niagara and other falls are used for the generation of hydel power. The fresh water lakes are used for supplying water for agricultural land. The inland rivers, rivers of Asia, namely, Amudaria and Sirdaria discharge their water into the Aral Lake.

Thus we find the importance of lakes on the economic life of man.

Influence of Canals

Canals are excavated by man for various purposes. The canals are mostly used for (i) irrigation, (ii) transportation, (iii) pisiculture and (iv) other purposes. The canals of Punjab, U.P., Rajasthan etc. are extensively used for irrigating the agricultural land. These canals are the life-blood of these states. The canals developed from the river Nile have turned the desert area of Egypt into fertile agricultural land. The canals of France, Germany and Soviet Union play the vital role as inland waterway in these three countries. The canals of multipurpose river projects of India have great contribution in checking flood, soil erosion and developing irrigation system. Other than those mentioned above the artificial lake, natural water bodies, like tank, well, etc., also have great importance on the development of rural economy.

Influence of Climate

Climate as a factor of environment has far-reaching and wide and deep influence on man and his activities. It has direct and indirect influence on man. Climate directly influences man's energy, health, efficiency, food, clothing, shelter, etc. Again by directly influencing vegetative and animal lives, climate indirectly influences man's activity in utilizing these natural resources.

The influence of climate is more effective on the following :

Distribution of population and density of population depends much on climate. The polar region with very cold climate is unsuitable for human settlement whereas the temperate region with moderate climate is highly suited to man's settlement. The equatorial region with heavy rain and high temperature throughout the year is not ideal for human settlement. Hot desert is thinly populated.

Basic needs of man are food, clothing and shelter. Food is mainly obtained from staple crop. Different food crops are grown under different climatic conditions. Wheat is the product of the temperate region whereas rice is principally the crop of the monsoon region. In hot lands fish is the principal protein food but in the temperate climate meat is a principal diet along with cereal food. In the polar region meat and fish are the principal food.

For protecting the body civilized people use various types of garments. In hot climate cotton cloth is common. In cold climate wollen garments are essential. In desert area a particular type of garments are used.

The type of shelter and the materials used for its construction depend much on climate. Bamboo huts and sloping roofs thatched with straw are commonly used in the equatorial region. In polar region ice-built house is used. In developed temperate and tropical areas brick-built houses are common. In the desert region special type of house is constructed.

Economic activities: Economic activities of man are also influenced by climate. In tropical grassland hunting is an important occupation of man. Monsoon climate is highly suited for agriculture. Mediterranean region is noted for fruit-farming. Temperate region is noted for mechanised type of agriculture and industry. The development of industry and transport system in any region is influenced by climate. Climate influences industry directly and indirectly. The raw materials for the industry are obtained from various sources. Raw materials from forest products, agricultural products and animal products are the contribution of climatic condition in any region. Energy and ability of industrial labour is influenced by climate. Production of hydel power depends much on climatic condition.

Transport system in any region is affected by strong winds, heavy rain and flood, snowfall and freezing of river, seas, and accumulation of ice and dense fogs. Heavy monsoon rain and melting of ice in the Himalayan region of India cause flood dislocating road, railway and river transport. Northern and North-Eastern parts of Canada during winter months become ice-bound making them unsuitable for use. Air transport is affected by bad weather condition. Cultural development of any region is also influenced by climate. Architecture has followed the dictates of bright, sunny Mediterranean climate

Influence of Soil

Soils have great influence on plant life. Without soil, agriculture is not possible. Population is dense where the soil is fertile. Soil forms the basic need of agriculture. Soil gives support to plants and supplies moisture and food to the plants through their roots. Different types of soil contain different types of plant food and for different types of crops different types of soil are necessary. As for example, rice requires clayey alluvial soils, cotton grows well in black sandy loams, tea thrives well in red soils rich in iron and nitrogen. Infertile soil is unsuitable for agriculture. Of course man today increases the fertility of soil by adding manures and fertilizers and dry sandy desert soil is used through irrigation. Soil fertility is also increased by rotation of crop, keeping land fallow for a period after using it for sometime.

Influence of Animals: Animal has great effect on man's occupation. Animal is used as a beast of burden, and for drawing cart and plough. Yak in Tibet, lama in Bolivia, camel in Sahara are extensively used. Animal supplies a great variety of products such as milk, meat, bones, wool, silk, fur, feathers, hides and skins, etc. Animal-hunting. Stock raising, dairy farming, sericulture, etc., are the important occupations of man.

The people of Tundra Region, the Eskiomoes, depend very much on animal—the reindeer. They drink its milk, eat its meat, make tents for living with its skin and its tendrons are used for making ropes. Utensils and instruments are made of its bones.

Of course, today, in the days of mechanisation, use of animals in transportation is gradually declining.

Influence of Vegetation: Vegetation has great influence on man's life. Vegetative life purifies the atmosphere by drawing carbon dioxide and giving out Oxygen.

The Vegetation can broadly be classified into three groups, namely, (i) Forests, (ii) Grasslands and (iii) Shrubs.

(i) Different types of forests are found in different parts of the world. Various types of benefits are derived from forests by man. Even today in some areas forests are the sources of food, shelter and even clothing.

The benefits we derive from the forests are the following:—

- Various types of raw materials such as timber, fruits, rubber, cork, camphor, gum, resin, turpentine, creepers for medicinal purposes, bamboos, etc., are obtained from the forests.
- Some industries depend much on forest products. Soft timber and grasses supply raw materials for paper pulp. Match-boxes, match sticks and packing boxes are produced from soft timber. For making furniture and for other constructional purposes timber is extensively used. Lumbering industry is highly developed near the temperate forest of Canada.
- Forests make the atmosphere humid and thus cause rain. Flood and soil erosion, stormy wind, etc., are checked by forests.
- Forests supply food to man and animal. Fruits are collected for man's use. Roots and leaves are fed by an animals like cattle, sheep and goat.

But in some parts forests stand in the way in developing the region economically. It is very difficult to clear the evergreen forest areas of the ever hot and ever wet equatorial regions of Congo (Zire) and Amazon basins for agricultural purpose. The climate becomes unhealthy for human settlement and development of transport system becomes difficult.

Many areas are occupied by grasslands. Grasslands can broadly be divided into two groups—tropical and temperate. The tropical grasslands are the home of both herbivorous and carnivorous animals. Hunting is the principal occupation there. On the other hand, temperate grasslands of the U.S.A., Canada, Argentina, Australia, etc., are the principal agricultural regions for temperate crops like wheat, barley, etc. Livestock rearing is an important occupation of the people. Fertile soil of temperate grasslands rich in nitrogen is ideal for wheat.

Scrublands and shrublands supply fodder for animals. These grasses prevent soil erosion and approach of desert condition.

Influence of Minerals

Minerals are the backbone of modern industry. The standard of economic development of a country is measured how efficiently it could utilise the mineral resources. Mineral is also the index of civilisation. The gradual change from stone age to atomic age gives an idea about the trend of efficiency and enterprise man could achieve in developing economy. Coal, petroleum, atomic energy supply power to modern industry and transport. Iron, copper, lead, tin, aluminium supply raw materials for metallurgical industries. The highly developed industrial countries of the world like the U.S.A., U. K., U. S. S. R., W. Germany, France, Japan, China are the great consumers of minerals. India's industrial progress after independence is due much to the production and utilisation of her minerals.

Minerals are utilised for the following purposes :

(1) Minerals are used in industry as raw materials and sources of power. Iron and steel, Aluminium, metallurgical industries consume iron-ore, Bauxite, etc., as raw materials and coal and mineral oil as sources of power. Industrial activity of the U. S. A. and U. S. S. R. depends on richness in their minerals.

(2) Modern transport system namely, railways, automo-biles, aeroplanes, ships, etc., depend much on iron, aluminium, tin, etc. Coal and petroleum are used as power. Large quantities of aluminium, copper, etc., are used in electrical industry.

(3) Agricultural machineries are produced with the help of minerals and manures are obtained from potash, phosphates and nitrates.

(4) Minerals are the great sources of foreign exchange. Oil-field of the Middle Eastern countries and iron-ore of India are the great earners of foreign exchange.

(5) Gold and other metals are used for making coins. Minerals are used for making ornaments, jewellery, household utensils, medical and surgical instruments.

But the minerals are not uniformly distributed. So if minerals are not properly obtained economic-activity is retarded. In the political world the minerals have great influence. Arab countries control over oil has great influence on world's trade and politics.

Of the different non-physical factors of environment, Population, and Political and Social Organisations are the most important.

Influence of Population

Population of my country plays the outstanding role in the economic development and acts as the principal resource in shaping the economy of a country. A man with his power of invention, physical and mental energy, culture and quality utilises the resources available under different environmental condition to develop economy. Without his help no activity is possible. A country may be otherwise full of varieties of resources. Unless these are properly utilised by man, no development is possible.

There are many countries which could not be developed properly on account of scarcity of population. The countries like China, India could develop agriculture as a large number of agricultural people are there. The U.S. A., U. K., Japan, etc., were developed in agriculture, industry, trade and commerce by a large umber of people. But a large number is not always the only ideal condition for the economic development. High-level manpower is needed. The advanced countries like the U. S. A., U. K., U. S. S. R., etc., have continuous pressure to innovate and make new discoveries in Science and Technology. The resources are utilized by the introduction of new techniques.

There are many countries in Asia, Africa and South America where the number of people is more than what they can accommodate. But in comparison with the industrially developed countries of the world, the density of population is not high. These countries are over-populated because they could not make the best utilisation of the economic resources. Because of the low level of economic development the standard of living is very low. But by the proper utilisation of economic resources better living facilities can be ensured and economically these countries can be made well developed and the problem of population will be minimized. India, through proper planning and economic development in different sectors can only make the country rich and solve the population problem.

So the number, the quality and culture of the people matter much in the economic development of any region.

Influence of Political and Social Organisations

Political and social organisations of a country have profound influence on the economic life of a country. Though a state has full responsibility to maintain law and order, to ensure security , to promote social welfare through education and care of health of the people yet it has a significant role to play for its economic growth. The socialistic pattern of society which India desires to achieve has become realistic by the adoption of economic planning through public and private sectors. India is heading towards industrialisation and within a few years will be able to solve the food problem which is really an acute problem for the country. A stable popular Government can only shape the economy of a country on planned basis spread over a long period. Efforts on the part of India to improve the conditions of the people through peaceful co-existence with other economically developed countries of the world are really sound and fruitful.

By establishing trade relations and borrowing technical and financial help from the developed countries, an undeveloped country can improve its economy.

Social organisations also contribute much towards the economic prosperity of a country. The joint family system, high birth rate, etc., stand in the way of India's economic progress.

Adaptation of Man to his Environment

Different groups of people living in different parts of the earth differ from one another in respect to their physical and mental characteristics, food, dresses, languages, religion, race, mode of living and above all inherent culture and tradition. Of these, some are, such as body feature, colour, race, religion, etc., hereditary but others are controlled by physical and cultural environments. Some people through their inherent skill and quality, culture and tradition, education and scientific achievement have been able to make their mode of living very high and enterprising and others are even today uncivilised and carry on primitive mode of living surrounded and influenced by the local environmental conditions. As for example, people of the Western

European countries, Soviet Union, United States, Japan, China and India are civilized and advanced group of people. They have developed their agriculture, mining, industry, trade and commerce very greatly. Scientific development and invention have made their life more comfortable and advanced. On the other hand, Pygmies of Central Africa, Bedouins of Arabic and North Africa, the Eskimoes of North America, Samoyed of the U. S. S. R. still live a primitive life depending primarily with local environmental conditions. Though they have come in touch with civilized group of people, yet the influence is very limited and localised.

Though the mode of living of the different groups of people have been profoundly influenced by the factors of environment-physical and non-physical, yet the civilized groups of people have been able to control and modify the factors of environment in their favour. On the other hand underdeveloped and uncivilized groups of people are still the slaves of nature. The way a group of people try to conquer and overcome the difficulties put forward by their environment through their inherent skill, culture and scientific achievement is called the *adaptation of man* to his environment.

Influence of Environment and Adaptation of Different Groups of People

To meet the daily needs, namely, food and drink, shelter, dresses, tools, transport and other amenities of life under dif-ferent enviornmental conditions endowed with different types of culture and quality carry on different types of occupation. How do these occupations vary from region to region with the corresponding change in environmental and cultural conditions are *described below in a nutshell:—*

A. *Primitive occupations,* namely, *Gathering, Hunting* and *fishing:* This type of occupations are undertaken primarily by the primitive people. Their are still some areas where this type of occupation is carried on.

The *Tundra Region of North America* and *Asia, the Northern Mexico, the Congo basin of Africa and Amazon basin* of *South America, South-West Africa* and *Bachuanaland* and in some parts of *Australasia* still have some primitive groups of people who preserve their primitive life.

The Tundra region is occupied by perpetual snow and ice. Agriculture is difficult. The Eskimos and Samoyeds rear Raindeer, live in special type of dwellings made of ice and skin, carry on hunting of animals and catch from the sea seal, whale and other fish. Civilized group of people in North America and Asia are trying to develop the region.

In Congo and Amazon basins ever hot and ever wet climate help the region for the luxuriant growth of evergreen hard-wood forests. Agriculture is difficult. People generally live on tree top and collections of fruits, rubber, and hunting are the principal occupations. Of course advanced group of people have cleared the forest margin for agriculture—specially for growing rubber, rice and sugar-cane.

B. Pastoral Nomadism : The Tropical and Temperate grasslands where the rainfall is very small, agriculture is not possible. Animal rearing and hunting are the most important occupation of the people. The nomads move from place to place with their herd of animals. *Kirghiz* of Central Asia, *Masai* of East-African Savannah, *Kikuju* of Northern Kenya, and *Navaho* of Mexico and Arizona are important primitive groups of people who carry on these types of occupation.

C. Forest Dwellers carrying on *Hunting. Trapping* and Shifting *Cultivation : Equatorial* and *Tropical evergreen forests* of Congo and Amazon basins, Indonesia and forests of South-East mountain areas have hot and wet climate. Forest growth and growth of weeds and creepers very quick. Clearing of forest is difficult. Agriculture is carried on by the primitive people by shifting method. Hunting and trapping are the most important occupation of the people.

In the landforms forests of North America and Eurasia, Agriculture is carried on in limited areas. Lumbering is the most important industry. Far-bearing animals are caught by the people. Far-trading is an important occupation of the people. Trees are cut during winter months and they are either dragged when thawing takes place or through rivers during summer months. Valuable soft timber are highly useful for paper pulp, match boxes, match sticks, furniture etc.

D *Intensive subsistence type of Cultivation of Monsoon lands:* In South East Asia, especially in China, India, Java and Japan rice is cultivated by intensive subsistence type of farming. Ganges Valley & Delta and coastal plain of India, the Irrawady Valley of Burma, Yangsi-Kiang and Hoang Ho Valleys of China, the river valleys of Thailand and Indo-China are the principal areas of production. Other than rice, wheat, maize, sugarcane, and oilseeds are also grown.

E. *Plantation farming in Tropical Areas :* The white people during the 18th century developed the plantation farming industry with their capital and organisation and local labour in colonies. The principal of them are Rubber plantation of Indonesia, Sugarcane cultivation in

Java and Cuba; Coffee cultivation of Brazil and India and Tea cultivation of India, Sri Lanka and Eastern African countries. The soil and climate are very ideal for the production of these crops in respective areas. Of course the industry is now carried on by local people. These crops are very important earner of foreign exchange.

F: *Mediterranean type of Agriculture:* Summer hot and dry climate and winter wet climate help much in the production of various types of fruits, and wheat, rye, barley, etc. in the mediterranean lands. The principal fruit trees are, citrus fruits (such as oranges, lemons, grapes); deciduous fruits (such as peaches, grapes, the olive, almond and pig).

Olive is the distinctive crop of the mediterranean region. Where there is mediterranean type of climate, there is olive. Production of mulberry trees are also the important product of the region.

The lands around the mediterranean land, the valley of California, Central Chile, South-West and South-East Australia carry on producing mediterranean crops.

G. *Intensive mechanised type of crop :* This type of method is adopted in the densely populated regions of Western Europe and Soviet Union and in the temperate grasslands region of the world for the cultivation of wheat, sugar-beet, etc. Nile delta produces rice by intensive method of cultivation.

H. *Animal Rearing:* The tropical and temperate grasslands are used for rearing cattle and sheep both for meat, hides and skin, wool and milk.

Man's Impact on Changing Environment

The vital role of man in ecology has been stressed within several topics considered so far. This last chapter summarizes man's past impact on British ecosystems and presents two detailed regional examples. The ecological activities of man are a containing phenomenon and present and probable future patterns of these are also examined in ecosystem terms. The difficulties arising from multiple land use and mounting recreational pressures on our countryside are then considered, particularly with respect to the conservation problems they pose for environmental planners.

The Legacy of Human Interference

From about 5000 B.C. onwards man has had an ever-increasing influence in determining the distribution and many of the properties of our flora, fauna and soils. The early cultures, Palaeolithic, existing as hunters and gatherers at low population densities, were once thought to have produced only slight changes in their environment. However, for certain unpland habitats, such as the North York Moors and Dartmoor, Simmons has argued for a clearly detectable. Mesolithic impact and there is increasing evidence for this view. There was also impact on those marginal lowland sites which were ecologically susceptible to change by virture of their pedological characteristics.

By about 3000 B.C. Neolithic man had arrived in this country and his influence was certainly pronounced. As a cultivator, he cleared mole forest and to protect his own grazing animals and increase his food supplied he began to reduce the competing native fauna of the woodlands. We have strong evidence that selective clearance of some tree species was taking place at this time (young lime and elm trees to supplement scarce grass supplies for domesticated cattle) and some slopes were experiencing soil erosion as a result of use by man. The first Neolithic forest clearances were temporary plots which were subsequently abandoned reverting by subseres towards the climax forest composition. Before long, however, more permanent clearances were established and by then much of the forest in some regions would have been essentially secondary in form. This was particularly so on the better drained soils and slopes. It was initially thought that a high concentration of prehistoric remains on limestone scarplands (*e.g.*, the Chilterns and North Downs) and uplands (*e.g.* Dartmoor, Pennines and Forth York Moors) meant that early man avoided lowland sites. But recently many more prehistoric settlement sites have come to light in the lowlands causing a revision of this view. During the following Bronze Age and Iron Age the forest underwent further reduction–Bronze Age settlements, for example, reached quite high levels on Dartmoor.

It was these cultures that the Romans encountered during their invasion of Britian. In this conquest they set fire to woodland as part of their military strategem against local populations. The Romans were also important in clearing lowland areas for cultivation. They were able to tackle some sites which would have proved unattractive to earlier cultures whose technology for dealing with the denser vegetation and poor drainage of the wetland valley bottoms was limited. Throughout this and earlier periods the British flora was receiving additional species and Godwin lists many introductions which are now largely accepted as native. These include Castanea sativa (sweet chestnut), Papaver rhoeas (poppy), Juglans regia (walnut) and Sinapis arvenis (wild mustard).

Forest reduction must have continued throughout the Roman period and into the Dark Ages but there would still have been vast tracts of woodland landscape. The next clear evidence of profound environmental alteration comes with the arrival of the Saxons. From about the sixth century onwards, these agricultural colonizers opened up much of the lowland forest and established perhaps 90 percent of the existing pattern of English villages and hamlets. The process used was 'assarting' or grubbing up and burning trees and shrubs to prepare the land for tillage. About the same time, Norse, Danes, and Viking first raided and later settled many parts of nothern Britain. They, too, used fire extensively to establish their rule. Under the Norman feudal system, the clearance and reclamation of land continued, but large areas were also set aside as royal preserves—open 'wastes' and woods suitable for game hunting, such as the New Forest. However, their introduction of the rabbit and their fostering of deer herds did much to restrict the regeneration of trees.

These destructive trends gathered pace during the Medieval period and locally timber supplies became very scares. The Exchequer Rolls of Scotland show Baltic timber being imported to Edinburgh as early as A.D. 1329. By the seventeen century, Samuel Pepys, John Evelyn and other prominent writers were bewailing the dearth of timber in Britain. Much in the south had been used for fuel, buildings, ship construction and charcoal production. Increasingly, merchants moved into remote regions for their supplies (*e.g.*, the Lake District, the Loch Lomond Woods). On Roman maps the simple descriptive term Coledonia Silva—the Caledonian Forest-had been boldly written across Scotland, yet in A.D. 1617 Sir Anthony Weldon, touring lowland Scotland, was led to the comment, 'there's scarce a tree to hang a Judas on, and indication of the completeness of tree removal in these parts.

Regional Examples

The human impact on the natural environment can be followed more specifically if we take two detailed examples. The first concerns the pine and birch woods flanking the Cairngorm Mountains of Deeside and Speyside. They are remnants of the ancient Caledonian Forest. In complete contrast are the numerous small mixed-deciduous woodland plots lying a few km east of Leicester, many of which are fox coverts. Some of these are much altered in character but originally were part of the once extensive Summer Deciduous Forest. Others, however, were entirely created by man as planted woodland for fox-hunting. Where are most present coniferous woodland in Britain in quite obviously planted, so much of the deciduous woodland is equally ' artificial'.

The Cairngorm Forests

These forests formerly extended to altitudes of 620 m or more but today the upper limit seldom exceeds 490 m. Much of the forest is now open ground or forestry plantation. The gross disturbance of these woods began about A.D. 400 in Pictish times when large tracts of lower forest were cleared for primitive cultivation. The decline of forest on the higher slopes took place much later, mainly in the Medieval period, with the evolution of cattle-breeding, the main-stay of the Highland economy until the late eighteenth century. Gaffney has shown that by the sixteenth century most forested land between 300 m and 600 m had been turned into highly prized grazings, Through the area ran several important drove roads for moving the cattle to summer pastures (shealings) or to lowland markets. These routes attracted rivers or thieves who were extremely active until the military set up control points in the region. In addition to the local cattle on the summer pastures, there were ' strangers' or gall cattle. For local tenants it was common practice to take in such cattle for a fee from farms as remote as 35 km away. We cannot be certain just how well wooded the higher ground originally was but there are numerous references in the Gordan Castle Estate Papers to unlawful grazing and heavy exploitation timber at these levels. It is clear, however, that the trend was always towards continual reduction of the forest cover.

Other animals, in addition to cattle, featured in the economy and ecology of the pre-sheep period of this part of Highland Scotland. Deer were numerous and greatly encouraged by the Dukes of Gordon, the upper glens eventually becoming over-stocked. In winter their traditional feeding grounds were the native pinewoods. Horses, goats and sheep were also numerous. Macfarlane in 1748 described the upper Dee as richlystocked with these animals. In 1779 the Forest of Glenavon was 'eaten up like a sheep pasture' and in 1794 'the forest contained many green spots and...affords pasture for a thousand head of cattle'. It is obvious that these animals would have a marked effect on tree seedling regeneration and ground flora composition.

Several activities associated with called husbandry also had a pronounced influence on the woodlands. First, the construction of temporary bothies ('scalans') at the highest pastures required the use of 'divots and trees'. Secondly, he low, cattle had to be confined or 'hained' at night and this often necessitated the building of timber enclosures. Thirdly, crops needed protection from grazing animals and this was often achieved by erecting timber barriers ('gathering'): at one site 5,000 young trees were used to protect a cornyard. A measure of this destruction of native trees is given by the petition of James Grant of Grant to the Parliament of James VI 'that fir wood in Abernethic and Glencherneck are dayly cutt, stollen and carried away by the tenants of Strathspey... and that the birk-woods are wolly destroyed by peiling of the timber peiled standing rotting in the woods... and against the great hurt and prejudice done to the fir woods of Strathspey by cutting and destroying standing trees for to be candle fir to all the inhabitants'. (During times of hardship peeled birch bark was used as a human food source. The 'fir' mentioned is Scots pine, candle fir being a primitive form of home lighting.)

The practice of muirburning was of even greater ecocogical importance. This was widespread and frequently the fires got of control. It because so excessive that by 1695 severe penalties were enforced for burning too near the pinewoods—a third offence merited hanging. By the end of the eighteenth century large-scale sheep farming had replaced cattle farming but grazing and muirburning continued as dominant activities working against the remaining forest over. In the Cairngorms sheep were never so important as in other parts of the Highlands and eventually most sheep runs were converted to grouse moors. Again, this meant the continuance of regular muirburning.

The eighteenth century also saw the intensive commercial exploitation of the remaining woods by the timber companies, The York Building Company of Hull was responsible for the main onslaught in Abernethy Forest from 1728 onwards. Similary, Glenmore Forest was worked over in 1783 and the adjacent Rothiemurchus Forest at this time was proving $10,000-$20,000 annual timber revenue. Several severe forest fires occurred in these woodlands during the eighteenth and nineteenth centuries and there was further jheavy timber exploitation in the World Wars of the twentieth century.

The nobility have acted unwittingly to preserve some parts of these forests through their strict ownership and sporting interests. In the nineteenth century there was extensive tree planting by the nobility and one Gairngorm parish could boast 14 million 'firs' planted by 1877. However, the main plantings seen today are due to the Forestry Commission who, from 1923 onwards, began a comprehensive re-afforestation programme using several exotic conifers (mainly spruce, lodgepole pine, larch and fir). Glenmore became a National Forest Park in 1948 and in 1954 the large Cairngorms National Nature Reserve was declared.

Many of these recent measures to repair the ravages of previous centuries are now placed in jeopardy by the current massive influx of tourists, a phenomenon which we shall examine shortly. Recreational and sporting activities, forestry, conservation, and grazing, all operating together at these altitudes must lead to a severe clash of interests which will have to be resolved successfully if these already much modified woodlands are to survive in an acceptable form.

The Woodlands of East Leicestershire

The subdued, rolling landscape of East Leicestershire (which now included the former country of Rutland) stands in sharp contrast to the grandeur of the Cairngorm Mountains. The East Midlands are an important agricultural region but, nevertheless, the area contains a surprising number of small scattered woodland plots: 130 are shown on the O.S. 1:25,000 Leicester (East) SK 60/70 sheet. Many of these are less than 200 years old. Nearly all owe their form and existence to the fox-hunting activities for which the country is famous: the natural Summer Deciduous Forest which clothed the region in prehistoric times has long since given way to farming.

In the twelfth century all of Leicestershire and neighbouring Rutland was declared 'forest'. *i.e.*, wooded and open waterland reserved for hunting. Leicester became exempt in A.D. 1235 but Rutland did not until much later . Some of the present-day woodlands around the parish of withcote, although now much altered by man, may be directly descended from the extensive early Medieval Leighfield Forest that was retained in this part of the Midlands until relatively late. The largest of them, Owston Wood, once grew very fine native oaks but these were clear-felled after the First World War. This wood was neglected for 35 years, acting simply as a source of stakes and fence material for the Cottensmore Hunt and local farmers. In 1960 the Forestry Commission took it over but have had continuing problems with weed species (willow herb, bracken, bramble and, particularly, willows) that became well established during the period of neglect. Another difficulty is that the heavy Lias Clay soils puddle badly in winter and crack deeply in summer, this condition being made much worse by the constant churning of the Cottesmore Hunt who draw these woods each winter.

Dillesdon Coplow is even more of an artefact since much of the hill consists of mature planted beech reaching 30 m or more. The wood also has a larch plantation and open areas are planted but with a mixture of Scots pine and beech. A few of the original oaks remain. Botany Bay Fox Covert, adjacent to Dillesdon Coplow, was originally planted soon after the time of the Australian settlement, as the name suggests. It is mainly degenerate oak-ash wood with an open canopy which has encourged profuse growth of hawthorn, hazel, blackthorn, privet and elder shrubs. The hazel was formerly coppiced but some shrubs are now 'chopped' to promote a tangle under growth suitable for fox cover.

Knighton Spinney, now surrounded entirely by suburbs lies 4.5 km from the centre of Leicester. It is an entirely man-made woodland and many other examples like it could be quoted from the East Midlands. Formerly the site was ploughed land but in the early nineteenth century it was planted with rows of oak. Ash, elm, sycamore and willow are now also present. The wood is rather neglected and has a poorly developed ground flora. It is no longer drawn by the Fernie Hunt and now forms the core of a municipal park in the Leicester suburbs.

In the late nineteenth century, at the height of fox-hunting as an acceptable and overt social phenomenon, over 300 horses might leave centres such as Market Harborough or Melten Mowbray for a day's sport in these woodlands no. Today, despite antagonism from some sections of society (a topic discussed later in this chapter), fox-hunting has not diminished and there are more hunts than at the turn of the century. In a landscape which is predominantly agricultural, it will be interesting to see what happens to these small but important woodlands in the future if rapidly rising costs and the present pace of social change leads to a decline in fox-hunting. Already many are under increasing pressure from tourists, and in several, vandalism is evident. Like the remote Cairngorms, these woodlands represent a dynamic situation. In both cases there is overwhelming evidence of man-induced changes during the past and a certainty for the future that man will become more not less important in the ecology of these areas. These two regional studies could be matched by similar examples drawn from virtually any part of the British Isles. As we shall see in a later section, for all British woodlands the basic problem is to so manage this resource that the pressures emanating from the various human uses do not undermine good conservation practices.

Much that we cherish in the countryside today is artificial and results from centuries of environmental modification by man. The open hillsides of the Lake District, the Chalk Down-lands with their distinctive chalk floras, the expanses of neither moor and the 'typical' rolling English countryside of fields, hedges and small woodlands all fall into this category. Some species have proved resilient to change and others have adapted well, extending their range into man-made environments. The seagull, for example, appears to be just as much at home scavenging from inland urban refuse tips and frequenting reservoirs as it does in its more usual costal habitat. Likewise, pigeons, closely related to rock doves, find suitable nesting sites on the 'cliff ledges' of office buildings. An even more interesting case is the collared dove (Streptopelia decaocta). This native of India reached Turkey in the sixteenth century but has now shown an explosive spread across Europe during this century. First reported in Lincolnshire in 1952 it had reached a population of 30-50,000 in Britain by 1970. Coombs *et al.* report that it is invariably associated with human settlement, particularly favouring town parks and suburban gardens. The introduced Oxford ragwort (Senecio squalidus) is an example from the plant world. It has spread rapidly in Britain along the cinder-strewn embankments of railway and into other disturbed stony sites where conditions approximate to the lava fields around Mount Etna, one of its original habitats in Sicily. An increasing number of species, however, are highly vulnerable to these changed landscapes and some have become extinct while many are now rare. These trends must be carefully considered as we contemplate present and future patterns of activity in rural Britain.

Present and Future Impact

The increasing pressure of recreational pursuits on the countryside was previously mentioned and this will now be examined in more detail. The developments described mainly relate to the Cairngorm Mountains but most of the findings are of general application because many other parts of the British Isles are undergoing a similar experience, *e.g.* the Lake District Dartmoor and the Peak District.

In the Cairngorms the hill-walker and mountain-climber, always keenly appreciative of what the region offers, have now been joined by rapidly growing numbers of motorised tourists. For some time tourists had limited access to the region but now there have been road improvements and a great extension of tourist facilities. The small settlement of Glenmore has been expanded and the road widened and continued to 760 m. Chair-lifts, ski-tows, snack bars and restaurants now exist at the terminus. These early investments soon bore fruit. On Easter Monday 1962, 5,000 people were estimated to be on Cairngorm itself and during the first seven months of operation 80,000 people used the chair-lift. Over a five-day period in 1965 the Cairngorm Board collected some $4,500 on lift and toll charges. Commercial success like this prompts more investment, an example being the nearby $2 2/1 million Aviemore Centre which includes two large hotels (now frequently used as conference centres), chalets, a swimming pool, an ice-rink and artificial practice ski-slopes. In 1982 further proposals for expansion into adjacent corries were under consideration (including two ski-lifts, two tows, parking for 1,000 cars, cafeteria, toilets and snow fences). Lickorish, writing about planning for recreation and leisure, noted the point made by the US Department of Commerce as early as 1950: when a community can attract a couple dozen tourists daily throughout the year, it is economicaly comparable with acquiring a new manufacturing industry whose annual payroll is $100,000. And business interests soon realise that an investment of £100,000 in a mountain that brings in £10,000 profit a year is just as useful as the same money made from more conventional investments. However, some might see the situation rather differently in that such schemes may not create as many jobs as the same money more conventionally invested and therefore the end result might be less socially acceptable.

An increasing number of visitors to the Cairngorms are people on day trips. Many will come in their own vehicle but there is a rapidly growing bus and coach tour element in these numbers (this expansion in the early years of these recreational developments can be seen from Scotland's bus and coach tour passenger number which increased from 3,900,000 in 1960 to 5,800,000 in 1963). Because of the relative remoteness of many recreational areas this implies a lot of travelling which leads to congestion on roads originally built to serve small, isolated Highland communities.

In the context of the United Kingdom, Dower, writing on the function of open countryside, has considered some of the main conclusions of the 28- volume report published in 1962 by the Outdoor Recreation Resources Review Commission of North America. The main American findings were:

1. The population of the United States will double by AD 2000 but the national demand for outdoor recreation will treble.
2. The recreational areas are remote from the centres of population.
3. Driving for pleasure is today the most popular outdoor activity.
4. Most people try, to centre outdoor recreational activities around a body of water.

It is more than 20 years since Dower's paper and no such detailed comprehensive report on recreation exists for this country. However, an increasing number of specific areas within the British Isles have since been produced (*e.g.* Moss; O' Riordan) and several general texts, often viewing the recreational problem in global terms, contain much that is relevant to this country (*e.g.* Simmons; O' Riordan and Turner). It is now quite clear that the pattern of leisure-time activities in Britain has moved strongly in a similar direction to that seen in the United States. We already know that the majority of motorised tourists seldom venture more than a few hundred yards from their car and this often gives rise to the 'honey-pot phenomenon' where large numbers congregate at popular vantage points. A good example in the Cairngorms is Loch Morlich, where erosion of the shoreline is already a problem (similar shoreline erosion has been reported by Tivy and Rees for many Scottish lochs). There is a strong case for preserving extensive parts of the Cairngorms against massive penetration by tourists, but although a large Nature Reserve does exist the pleasure-seeking public have never shown great respect for lines on maps. Such a vast influx of tourists coupled with unplanned piecemeal development will do much to destroy the essential character of these mountains.

Problems of Multiple Land Use

There is agreement that development in regions such as the Cairngorms must be planned and should consider all interests, including the larger national interest. The problem is to find a plan broadly acceptable to all and, at the same time, capable of implementation at reasonable cost. Difficulties arise because interests clash and priorities have to be determined. One interest, that of the tourist, has been mentioned but others include those of the water boards, the power stations, the military, mining, forestry, agriculture, sporting activities, transportation and last, but not least, conservation. Many of these uses are mutually exclusive and nearly all have their problems of operation greatly increased by the presence of large numbers of tourists.

Another factor adding to the difficulties of resolving clashes of interest in our countryside is the question of land ownership. Much of our landscape is still privately owned and the land use pattern here will often, but not always, be determined by the necessity to make a profitable return on investment. Large areas are state-owned or state controlled and motives other than profitable return may then decide land use. But, it should be remembered, many designated National Parks do include farms and private land within their boundaries. This makes a comprehensive development plan for the whole region that much harder to achieve.

It is beyond the scope of this book to deal fully with each interest in turn. Instead, a few comments under each heading will be made so as to convey something of the complexity involved in multiple land-use planning. However, fuller treatment will be given to conservation because of its central role in preserving as a resource these very landscapes that we wish to use and because effective conservation calls into consideration many of the ecological principles outlined in previous chapters. This makes it of special interest to geographers.

Water Boards

The industrial and domestic demand for water has risen sharply in this century and it is usually argued that this will continue to be the case in the foreseeable future. For example, the installation of washing machines and dishwashers in the home and the rise in personal cleanliness have led to a dramatic increase in household consumption of water. Agriculture now uses more water and consumption is likely to expand even further in eastern England, where intensively grown crops are likely to benefit from the use of sprinklers and irrigation three years out of every five. Heavy demand, coupled with a very variable climate, already leads to local and regional shortages during spells of dry weather. However, lorecasting future demand for water is not without its difficulties. Several large traditional industries which are heavy users of water are now in marked decline (*e.g.* steel) Many of the newer industries based on electronic technologies will not require massive inputs of water. The problem is well illustrated by the Kielder Reservoir. Since completion of this recent £167 million scheme the industrial demand in the Northumbrian Water Authority area has already decline by 5 percent and full use of the water from Kielder is not now likely until the year 2000.

The creation of large reserviours must take place at topographically suitable sites which, by their very nature and location, will tend to be attractive areas for recreation. Many reservoirs have been sited in our uplands but an increasing number are now in lowland areas. Wherever the site is, local opposition to the loss of land involved can be passionate: a group opposed to a new reservoir on farm land in the Roadford Valley, west Devon, even threatened deliberate pollution of the River Tamar if the scheme went ahead. Extensive areas of farm or forestry land have been flooded by schemes and in the past the feat of water pollution and bank erosion has usually meant the exclusion of people from these sites. Fortunately, this policy has been revised of late and more reservoirs now cater jointly for water and recreational needs. This is well seen at the new Rutland Water Reservoir which lies in the heart of an important agricultural region. The loss of farmland here will be partly compensated for by the provision of recreational facilities, particularly for fishing, boating and water-sport interests but also providing picnic areas and nature reserves. However, at Cow Green, Upper Teesdale, plans to build a reservoir to serve the ICI industrial complex on Lower Teeside met great opposition. This was because the proposed site was the habitat of several rare plants, a unique arcticalpine floral element containing species such as Gentiana verna and Minuartia stricta which are confined in Britain to the Upper Teesdale area. Despite a well organized national campaign and strong support from many internationally eminent botanists the plans were implemented.

Another activity by water boards that threatens wildlife is the drainage of wetlands. Halvergate Marshes in Norfolk is a classic case where controversy between farming and conservation interests has raged for several years now. Drainage would allow profitable grain crops to be grown, but to the conservationist Halvergate is one of the last pieces of species-rice grazing marshland in Britain.

Military Training

The Ministry of Defence controls several large regions in the British Isles where the general public are excluded. Unfortunately, some of these regions are highly attractive landscapes with great recreational potential. Examples are western Dartmoor, parts of the Dorset coast and a portion of Breckland in Norfolk. Currently, these areas are under review with strong pressure for more to be released for enjoyment by the public. Some people have argued for an almost total reduction of military involvement in these regions. However, if we spend thousands of millions of pounds on defence it then seems illogical to deny proper training facilities. The importance of adequate preparation in the use of modern warfare technologies has been emphasized by the recent Falklands conflict.

In some cases the presence of the Military actually assists conservation. For example, access to most of the Pendine Sands of Carmarthenshire is prohibited because of a missile and gunnery testing range. Yet these dunes may well act as a refuge for native fauna and they do carry a fuller and more varied vegetation cover than adjacent sections which are regularly and heavily trampled by tourists. Likewise, the Chemical Defence Establishment at Porton Down in Wiltshire controls about 2,000 ha of chalk grassland which is now one of our most important wildlife conservation areas. It is a diverse region which acts as a sanctuary for many interesting species that are rare or absent in the surrounding zone of intensive agriculture.

Power Stations

By the very nature of their activities nuclear power stations must be sited at some distance from major settlements. The same applies to several of the associated Atomic Energy Authority Establishment (*e.g.* Dounreay, Windscale (now Sellafied) and Calder Hall). Because of this, many occupy areas that would normally be of prime conservational or recreational interest. This is particularly so since the vast majority of nuclear power stations also occupy coastal sites (*e.g.* the Central Electricity Generating Board's nuclear power stations at Dungeness, Bradwell, Sizewell, Hartlepool, Hunterston, Heysham, Wylfa and Hinckley Point). Not only are they massive structures in themselves but associated facilities and security considerations mean that a large area of coast is effectively prevented from being used for any other major purpose. Several of the possible sites for future pressurized water reactors (PWR's) of the Three Mile Island type are also coastal locations (*e.g.* Stakeness on the Moray Firth and Gwithian near Land's End).

Mining

Mining and quarrying are dirty, dangerous and unsightly operations and their intrusion into areas of natural beauty is to be deplored. They also generate a form of transport (the heavy truck) which is particularly damaging to the countryside and infuriating to other road users. Nevertheless, we need these raw materials for the very things we may value in other environments, such as good roads, important urban buildings, gravel driveways and garden rockeries. In the long term their worst effects on the landscape can be alleviated by insistence on proper soil restoration measures and the use of the flooded surface pits for recreational and conservational purposes. When carried out successfully, this can actually enhance the landscape. A good example is seen in the Cotswolds where some of the abandoned and now flooded, shallow, surface quarries are used as centres for fishing and boating, while others are reserved for wildlife. The complex will eventually cater for many needs in an area formerly lacking in surface water facilities. Another example of cooperation is seen at Thrislington, Co. Durham. Here, the Nature Conservancy Council and the Durham Conservation Trust are working in conjunction with a quarry company to move a 8 ha field complete with topsoil, rare plants (including several orchids) and animal life to a new site half a mile away. This will allow pure dolomite, an important ingredient in the steel industry, to continue to be mined.

Mining and quarrying are not always confined to hard rock areas, most of which occur as our uplands. Extensive lowland tracts have been worked for gravels, sands and brick clays, leaving numerous scars on the landscape. Only recently has the full recreational potential of some of these sites been appreciated. However, the picture is complicated because some lowland areas fall within National Parks. A case in point is the highly attractive Mawddach Estuary, which is near sea-level yet lies within the Snowdonia National Park. Because of this, plans to dredge for alluvial gold in the estuary aroused much protest.

Forestry

Many British woodlands are privately owned and contribute a very desirable element to our landscape. Whether they remain in their present form may well depend on government taxation policy as much as on anything else. In the past, governments have encouraged tree planting on private estates by a series of financial incentives. But any changes in capital transfer taxation following death of the owner could lead to a sharp decline in many of these woodlands. As table below shows, the private sector still produces nearly 60 percent of our cut timber and is responsible for the vast majority of our more attractive hard wood species.

***Table:** Volume of Timber Cut Annually in Britain: Values in Millions of Cubic Metres (Modified from Rooke, 1974)*

Source	*Softwoods*	*Hardwoods*	*Total*
Forestry Commission	1.55	0.04	1.59
Private sector	0.82	1.30	2.12
Total	2.37	1.34	3.17

The state owns vast areas of forest, mostly as coniferous plantations. Formed in 1919, the Forestry Commission was charged with the task of building up strategic reserves of timber following the depletion during the First World War. This has been a difficult task because of the long time-span for tree species to reach maturity and the poor growing conditions at many upland sites. The Second World War also caused further setbacks. However, the Forestry Commission is on target for its aim of 5 million acres (2 million ha) of afforested land (including both state-owned and privately owned). Nevertheless, we still import vast amounts of timber despite increasing home production. Wood and wood products are Britain's third largest import, costing about $ 2,370 million in 1978. Russia is a main supplier which could make our position vulnerable at times of international crisis.

Net imports are unlikely to drop much below 90 percent of our total timber consumption during this century.

Table: *Consumption, Production and Import of Timber in Britain at the Present and Expected for the Near Future: Values in Millions of Cubic Metres (From Rooke 1974).*

Year	*Expected Consumption*	*Home production*	*Expected net imports*	*Net imports as % of total*
1970	45	3.5	41.5	92
1975	51	4.5	46.5	91
1980	57	5.0	52.0	91
2000	82	8.7	73.0	89

One of the major difficulties with forestry as a land use is forecasting future trends within the industry because of the long-term nature of the crop. Raup has studied the basic concepts and assumptions underlying North American forest management and found them to be false. It looks as though a better assumption would be that predictions beyond one or two decades were more likely to be wrong than right' and 'uncertainties in the long run call for the greatest possible flexibility in resource use'. These lessons may have strong implications for British forestry. One of the major problems in Britain is the vast acreage now planted with fast-growing conifers, mainly Sitka spruce (Picea sitchensis). Sitka grown under our climate is largely a low-grade softwood timber and there is increasing difficulty in finding suitable and profitable outlets for it. While the need for more forest cover can be readily demonstrated, especially when given forecasts of a massive rise in world timber consumption and severe demands. being made on diminishing natural forests (see Bowman), technologies may change and so too may fashions for certain timbers. Thus, what we plant now may not be what is required some 50 or more years later. However, it is the tourist who is likely to cause the main changes in future forestry developments.

In early Forestry Commission plantations the public were discouraged, partly because of fire hazards and partly by the monotonous nature of the vegetation itself. But the key to present changes is contained in a declaration from the Seventh. World Forestry Congress (1972); 'Foresters recognise that forestry is concerned not with trees but with how trees can serve people...His allegiance is not to the resource but to the rational management of that resource in the long-term interests. of the community'. Forestry makes its greatest impact on the general public through the landscapes it creates. The Commission now realize that social and environmental demands must be harmonized with the need to produce timber. By 1974, the Commission had designated seven Forest Parks with a combined area of 176,600 ha and was investing over £1 million. annually in recreational provisions. The Commission had about 14 million day visits to its forests in 1980. In Holland, 75 percent of all state forests now cater for recreating and Britain is also moving rapidly in this direction.

Agriculture

Farming has often been in competition with forestry, particularly in upland areas where clashes also exist with some of the other forms of land use discussed. Indeed, as we have seen, most agricultural land has resulted from the clearance of forest at some time in the past. In 1972 there were some 11,000 hill and uplands farms in England and Wales. Much upland terrain is marginal for agriculture and may provide a better economic return and more rural employment under forestry. But the position is not simple, as James shows in his review; each case has to be carefully considered and conclusions applied to whole areas or even whole farms may be incorrect. Decision-makers have fallen into the trap of regarding agriculture in our uplands as a homogeneous activity when, in reality, some parts of an area or individual farm are more suited or less suited to agriculture than others.

We no longer live in a world of cheap food supplies (despite EEC surpluses) and a case can be made of either extending farming acreage or greatly intensifying productivity on land we already use. In lowland areas at least, intensive agriculture will increasingly involve large capital investments for complex technological system in farming–an aspect which at best makes the presence of too many tourists a nuisance and at worst a positive danger, according to Bonham-Carter. In some areas, however, the farming community has adopted a more welcoming attitude to the tourist. Along certain stretches of coast and in some unplands, farmers were not slow to realize that a field of caravans can be more profitable than one full of cows. On the flanks of Dartmoor a very large component in the income of some farms is now derived from activities associated with the tourist industry.

Government policy (*e.g.* guaranteed prices or the addition or removal of subsidies) and EEC decisions may well tip the economic balance of marginal agricultural areas one way or another, sometimes strengthening sectional interests and sometimes weakening them.

Schemes such as the drainage of wetlands may benefit the farmer, but seldom the taxpayer. Increased production is often in commodities already in surplus (*e.g.* milk, grains). This means that through the EEC's Common Agricultural Policy these surpluses are brought into intervention to maintain the price paid to the farmer (*i.e.*storage, until sold later and usually at a loss to the taxpayer). This largely explains the current crop of land use conflicts between farmers and others (mainly conservationist), *e.g.* Baddesley Common (Hampshire), West Sedgemoor (Somerset), Romney Marsh (Kent), Berwyn Mountains (Wales). The National Farmers' Union have responded by pointing out that an area the size of the Isle of Wight has been lost to British farming in the last 20 years and that 650,000 people are directly employed in agriculture and some 2 million (8 percent of the labour force) rely for their jobs on British farms and their produce.

Sporting Activities

In the countryside the fostering of sporting interests has played a large part in shaping the fine detail of the landscape. In addition to the numerous fox coverts in areas like Leicestershire, many plant communities were established in both upland and lowland districts specifically for game purposes, *e.g.* grouse moors, pheasant and partidge coverts. In some remote areas the revenue from sporting activities plays a significant part in the local rural economy. A recent survey of countryside sports shows that 4 million people go fishing, nearly 600,000 go shooting or stalking and 214,000 follow hounds, either on foot or in the saddle, at some time during the year—and these people come from a complete cross-section of the British public. These activities account for about £1,000 million of direct annual expenditure and employ either directly or indirectly about 90,000 people. However, not all look favourably on these activities. The 'blood' sports (involving hounds) are now under great threat. The Co-operative Wholesale Society, the largest independent farmland owner in Britan, has now banned fox hunting over its 12,000 ha and so have certain Labour-controlled councils. Such a ban was also included in the recent Labour Party election manifesto despite claims that hunting actually ensures the preservation of many woodland plots and the hunted species. Thus, as in most areas of land use conflict, a political factor must be recognized.

Catering for certain types of sporting interest is largely in private hands and takes place on private land. Good examples are grouse shooting and salmon fishing. However, there is an increasing development of sporting facilities of the general public and this is having its on our countryside. Today, large sums are spent on sporting equipment and there has been a great diversification in sporting pursuits. Although some activities, such as water-skiing or hang-gliding, appeal only to a minority, the clashes with other countryside users which will result from their expansion will soon have to be resolved. The creation of special countryside areas just for active sport is a legitimate use since these activities may do much to relieve the tensions and aggressiveness generated by sedentary or monotonous urban employment.

Transportation

For the foreseeable future we are mainly thinking in terms of road networks when transportation in the countryside is considered. In rural areas which attract many tourists the road network soon becomes heavily congested, having been established initially to meet only local needs. Any major improvement in the network may only result in more people flooding into the region. The better access the available may cause a greater strain on many of the other facilities in the district. So, to a certain extent, these improvements' can be self-defeating. For example, motorways recently constructed in southern England now give quick access from the populous south-east to the popular south-west and there is a danger of 'killing the goose that lays the golden egg'. A similar fear has been expressed for the Lake District in Cullinghworth's review of these problems.

The construction of major road systems means a loss of much land already in other uses. Gradient considerations dictate that this will often be the better quality agricultural land. By their very nature these major roads will usually follow the most open and obvious routes into a region and will be visible from many vantage points. While improving access from outside, they may cause local resentment and a breakdown of established patterns of social association for those living permanently within the region. The controversy over the proposed Aire Valley Trunk Road in Yorkshire stems from such local resentment. Here, a major planning enquiry was repeatedly disrupted by local pressure groups and finally had to be abandoned after costs of about £30,000 had been incurred.

Improvement of minor roads and trackways provides further access for tourists but may also bring benefits to agriculture. As Jones points out, one of the most serious physical handicaps to upland pasture improvement is the lack of good access roads. Millions of hectares are not fully developed because vehicles carrying lime and other fertilizers just cannot reach these sites.

Another major threat to rural areas is the construction of large airports. With their voracious appetite for land, their very specific site requirements and their attraction of numerous ancillary facilities and activities, they represent a major source of land use conflict. One has only to think of Maplin Sands and Stansted to be immediately reminded of these conflicts.

Having seen something of the problems that face the environmental planner we may now turn to a consideration of conservation, bearing in mind that each activity above causes special difficulties for the conservationist. When several are combined they may make good conservation practice virtually impossible: a neat exemplification of this is contained in the two well-illustrated papers recently produced under the general title 'Alarm call for the Broads', (Moss; O'Riordan).

Chapter 2

Basic Concepts of Ecosystem

An ecosystem is a community of living organisms (plants, animals and microbes) in conjunction with the nonliving components of their environment (things like air, water and mineral soil), interacting as a system. These biotic and abiotic components are regarded as linked together through nutrient cycles and energy flows. As ecosystems are defined by the network of interactions among organisms, and between organisms and their environment, they can come in any size but usually encompass specific, limited spaces (although some scientists say that the entire planet is an ecosystem).

Energy, water, nitrogen and soil minerals are other essential abiotic components of an ecosystem. The energy that flows through ecosystems is obtained primarily from the sun. It generally enters the system through photosynthesis, a process that also captures carbon from the atmosphere. By feeding on plants and on one another, animals play an important role in the movement of matter and energy through the system. They also influence the quantity of plant and microbial biomass present. By breaking down dead organic matter, decomposers release carbon back to the atmosphere and facilitate nutrient cycling by converting nutrients stored in dead biomass back to a form that can be readily used by plants and other microbes.

Ecosystems are controlled both by external and internal factors. External factors such as climate, the parent material which forms the soil and topography, control the overall structure of an ecosystem and the way things work within it, but are not themselves influenced by the ecosystem. Other external factors include time and potential biota. Ecosystems are dynamic entities—invariably, they are subject to periodic disturbances and are in the process of recovering from some past disturbance. Ecosystems in similar environments that are located in different parts of the world can end up doing things very differently simply because they have different pools of species present. The introduction of non-native species can cause substantial shifts in ecosystem function. Internal factors not only control ecosystem processes but are also controlled by them and are often subject to feedback loops. While the resource inputs are generally controlled by external processes like climate and parent material, the availability of these resources within the ecosystem is controlled by internal factors like decomposition, root competition or shading. Other internal factors include disturbance, succession and the types of species present. Although humans exist and operate within ecosystems, their cumulative effects are large enough to influence external factors like climate.

Biodiversity affects ecosystem function, as do the processes of disturbance and succession. Ecosystems provide a variety of goods and services upon which people depend; the principles of ecosystem management suggest that rather than managing individual species, natural resources should be managed at the level of the ecosystem itself. Classifying ecosystems into ecologically homogeneous units is an important step towards effective ecosystem management, but there is no single, agreed-upon way to do this.

History and Development

Arthur Tansley, a British ecologist, was the first person to use the term "ecosystem" in a published work. Tansley devised the concept to draw attention to the importance of transfers of materials between organisms and their environment. He later refined the term, describing it as "The whole system, ... including not only the organism-complex, but also the whole complex of physical factors forming what we call the environment". Tansley regarded ecosystems not simply as natural units, but as mental isolates. Tansley later defined the spatial extent of ecosystems using the term ecotope. G. Evelyn Hutchinson, a pioneering limnologist who was a contemporary of Tansley's, combined Charles Elton's ideas about trophic ecology with those of Russian geochemist Vladimir Vernadsky to suggest that mineral nutrient availability in a lake limited algal production which would, in turn, limit the abundance of animals that feed on algae. Raymond Lindeman took these ideas one step further to suggest that the flow of energy through a lake was the primary driver of the ecosystem. Hutchinson's students, brothers Howard T. Odum and Eugene P. Odum, further developed a "systems approach" to the study of ecosystems, allowing them to study the flow of energy and material through ecological systems.

Ecosystem Processes

Energy and carbon enter ecosystems through photosynthesis, are incorporated into living tissue, transferred to other organisms that feed on the living and dead plant matter, and eventually released through respiration. Most mineral nutrients, on the other hand, are recycled within ecosystems.

Ecosystems are controlled both by external and internal factors. External factors, also called state factors, control the overall structure of an ecosystem and the way things work within it, but are not themselves influenced by the ecosystem. The most important of these is climate. Climate determines the biome in which the ecosystem is embedded. Rainfall patterns and temperature seasonality determine the amount of water available to the ecosystem and the supply of energy available (by influencing photosynthesis). Parent material, the underlying geological material that gives rise to soils, determines the nature of the soils present, and influences the supply of mineral nutrients. Topography also controls ecosystem processes by affecting things like microclimate, soil development and the movement of water through a system. This may be the difference between the ecosystem present in wetland situated in a small depression on the landscape, and one present on an adjacent steep hillside.

Other external factors that play an important role in ecosystem functioning include time and potential biota. Ecosystems are dynamic entities—invariably, they are subject to periodic disturbances and are in the process of recovering from some past disturbance. Time plays a role in the development of soil from bare rock and the recovery of a community from disturbance. Similarly, the set of organisms that can potentially be present in an area can also have a major impact on ecosystems. Ecosystems in similar environments that are located in different parts of the world can end up doing things very differently simply because they have different pools of species present. The introduction of non-native species can cause substantial shifts in ecosystem function.

Unlike external factors, internal factors in ecosystems not only control ecosystem processes, but are also controlled by them. Consequently, they are often subject to feedback loops. While the resource inputs are generally controlled by external processes like climate and parent material, the availability of these resources within the ecosystem is controlled by internal factors like decomposition, root competition or shading. Other factors like disturbance, succession or the types of species present are also internal factors. Human activities are important in almost all ecosystems. Although humans exist and operate within ecosystems, their cumulative effects are large enough to influence external factors like climate.

Primary Production

Primary production is the production of organic matter from inorganic carbon sources. Overwhelmingly, this occurs through photosynthesis. The energy incorporated through this process supports life on earth, while the carbon makes up much of the organic matter in living and dead biomass, soil carbon and fossil fuels. It also drives the carbon cycle, which influences global climate via the greenhouse effect. Through the process of photosynthesis, plants capture energy from light and use it to combine carbon dioxide and water to produce carbohydrates and oxygen. The photosynthesis carried out by all the plants in an ecosystem is called the gross primary production (GPP). About 48–60% of the GPP is consumed in plant respiration. The remainder, that portion of GPP that is not used up by respiration, is known as the net primary production (NPP). Total photosynthesis is limited by a range of environmental factors. These include the amount of light available, the amount of leaf area a plant has to capture light (shading by other plants is a major limitation of photosynthesis), rate at which carbon dioxide can be supplied to the chloroplasts to support photosynthesis, the availability of water, and the availability of suitable temperatures for carrying out photosynthesis.

Energy Flow

The carbon and energy incorporated into plant tissues (net primary production) is either consumed by animals while the plant is alive, or it remains uneaten when the plant tissue dies and becomes detritus. In terrestrial ecosystems, roughly 90% of the NPP ends up being broken down by decomposers. The remainder is either consumed by animals while still alive and enters the plant-based trophic system, or it is consumed after it has died, and enters the detritus-based trophic system. In aquatic systems, the proportion of plant biomass that gets consumed by herbivores is much higher. In trophic systems photosynthetic organisms are the primary producers.

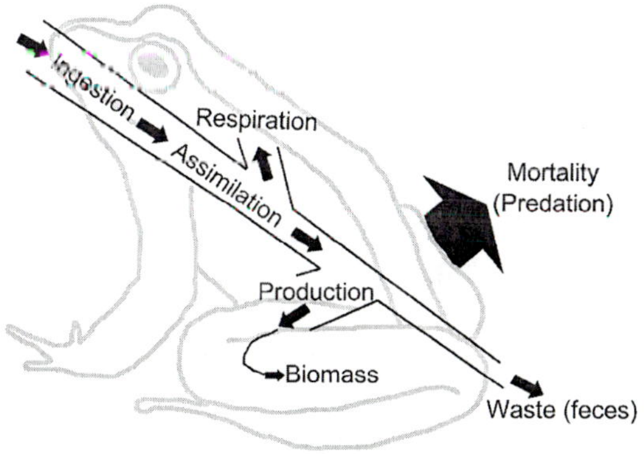

Figure: *Left: Energy flow diagram of a frog. The frog represents a node in an extended food web. The energy ingested is utilized for metabolic processes and transformed into biomass. The energy flow continues on its path if the frog is ingested by predators, parasites, or as a decaying carcass in soil. This energy flow diagram illustrates how energy is lost as it fuels the metabolic process that transforms the energy and nutrients into biomass.*

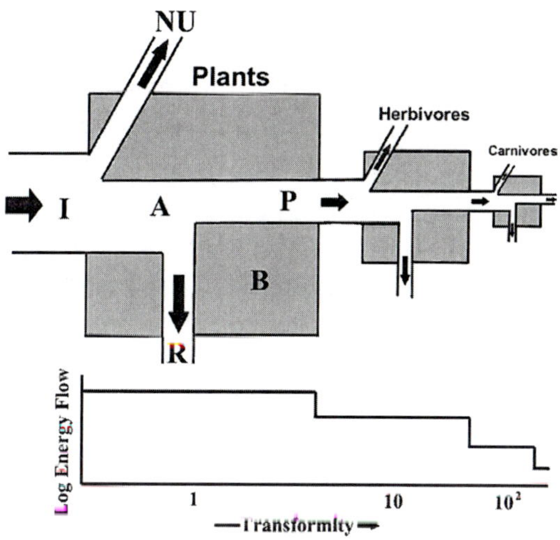

Figure: *Right. An expanded three link energy food chain (1. plants, 2. herbivores, 3. carnivores) illustrating the relationship between food flow diagrams and energy transformity. The transformity of energy becomes degraded, dispersed, and diminished from higher quality to lower quantity as the energy within a food chain flows from one trophic species into another. Abbreviations: I=input, A=assimilation, R=respiration, NU=not utilized, P=production, B=biomass.*

The organisms that consume their tissues are called primary consumers or secondary producers—herbivores. Organisms which feed on microbes (bacteria and fungi) are termed microbivores. Animals that feed on primary consumers—carnivores—are secondary consumers. Each of these constitutes a trophic level. The sequence of consumption—from plant to herbivore, to carnivore—forms a food chain. Real systems are much more complex than this—organisms will generally feed on more than one form of food, and may feed at more than one trophic level. Carnivores may capture some prey which are part of a plant-based trophic system and others that are part of a detritus-based trophic system (a bird that feeds both on herbivorous grasshoppers and earthworms, which consume detritus). Real systems, with all these complexities, form food webs rather than food chains.

Decomposition

The carbon and nutrients in dead organic matter are broken down by a group of processes known as decomposition. This releases nutrients that can then be re-used for plant and microbial production, and returns carbon dioxide to the atmosphere (or water) where it can be used for photosynthesis. In the absence of decomposition, dead organic matter would accumulate in an ecosystem and nutrients and atmospheric carbon dioxide would be depleted. Approximately 90% of terrestrial NPP goes directly from plant to decomposer.

Decomposition processes can be separated into three categories—leaching, fragmentation and chemical alteration of dead material. As water moves through dead organic matter, it dissolves and carries with it the water-soluble components. These are then taken up by organisms in the soil, react with mineral soil, or arc transported beyond the confines of the ecosystem (and are considered "lost" to it). Newly shed leaves and newly dead animals have high concentrations of water-soluble components, and include sugars, amino acids and mineral nutrients. Leaching is more important in wet environments, and much less important in dry ones.

Fragmentation processes break organic material into smaller pieces, exposing new surfaces for colonization by microbes. Freshly shed leaf litter may be inaccessible due to an outer layer of cuticle or bark, and cell contents are protected by a cell wall. Newly dead animals may be covered by an exoskeleton. Fragmentation processes, which break through these protective layers, accelerate the rate of microbial decomposition. Animals fragment detritus as they hunt for food, as does passage through the gut. Freeze-thaw cycles and cycles of wetting and drying also fragment dead material.

The chemical alteration of dead organic matter is primarily achieved through bacterial and fungal action. Fungal hyphae produce enzymes which can break through the tough outer structures surrounding dead plant material. They also produce enzymes which break down lignin, which allows to them access to both cell contents and to the nitrogen in the lignin. Fungi can transfer carbon and nitrogen through their hyphal networks and thus, unlike bacteria, are not dependent solely on locally available resources.

Decomposition rates vary among ecosystems. The rate of decomposition is governed by three sets of factors—the physical environment (temperature, moisture and soil properties), the quantity and quality of the dead material available to decomposers, and the nature of the microbial community itself. Temperature controls the rate of microbial respiration; the higher the temperature, the faster microbial decomposition occurs. It also affects soil moisture, which slows microbial growth and reduces leaching. Freeze-thaw cycles also affect decomposition—freezing temperatures kill soil microorganisms, which allows leaching to play a more important role in moving nutrients around. This can be especially important as the soil thaws in the Spring, creating a pulse of nutrients which become available.

Decomposition rates are low under very wet or very dry conditions. Decomposition rates are highest in wet, moist conditions with adequate levels of oxygen. Wet soils tend to become deficient in oxygen (this is especially true in wetlands), which slows microbial growth. In dry soils, decomposition slows as well, but bacteria continue to grow (albeit at a slower rate) even after soils become too dry to support plant growth. When the rains return and soils become wet, the osmotic gradient between the bacterial cells and the soil water causes the cells to gain water quickly. Under these conditions, many bacterial cells burst, releasing a pulse of nutrients. Decomposition rates also tend to be slower in acidic soils. Soils which are rich in clay minerals tend to have lower decomposition rates, and thus, higher levels of organic matter. The smaller particles of clay result in a larger surface area that can hold water. The higher the water content of a soil, the lower the oxygen content and consequently, the lower the rate of decomposition. Clay minerals also bind particles of organic material to their surface, making them less accessibly to microbes. Soil disturbance like tilling increase decomposition by increasing the amount of oxygen in the soil and by exposing new organic matter to soil microbes.

The quality and quantity of the material available to decomposers is another major factor that influences the rate of decomposition. Substances like sugars and amino acids decompose readily and are considered "labile". Cellulose and hemicellulose, which are broken down more slowly, are "moderately labile". Compounds which are more resistant to decay, like lignin or cutin, are considered "recalcitrant". Litter with a higher proportion of labile compounds decomposes much more rapidly than does litter with a higher proportion of recalcitrant material. Consequently, dead animals decompose more rapidly than dead leaves, which themselves decompose more rapidly than fallen branches. As organic material in the soil ages, its quality decreases. The more labile compounds decompose quickly, leaving and increasing proportion of recalcitrant material. Microbial cell walls also contain a recalcitrant materials like chitin, and these also accumulate as the microbes die, further reducing the quality of older soil organic matter.

Nutrient Cycling

Ecosystems continually exchange energy and carbon with the wider environment; mineral nutrients, on the other hand, are mostly cycled back and forth between plants, animals, microbes and the soil. Most nitrogen enters ecosystems through biological nitrogen fixation, is deposited through precipitation, dust, gases or is applied as fertilizer. Since most terrestrial ecosystems are nitrogen-limited, nitrogen cycling is an important control on ecosystem production. Until modern times, nitrogen fixation was the major source of nitrogen for ecosystems. Nitrogen fixing bacteria either live symbiotically with plants, or live freely in the soil. The energetic cost is high for plants which support nitrogen-fixing symbionts—as much as 25% of GPP when measured in controlled conditions. Many members of the legume plant family support nitrogen-fixing symbionts. Some cyanobacteria are also capable of nitrogen fixation. These are phototrophs, which carry out photosynthesis. Like other nitrogen-fixing bacteria, they can either be free-living or have symbiotic relationships with plants. Other sources of nitrogen include acid deposition produced through the combustion of fossil fuels, ammonia gas which evaporates from agricultural fields which have had fertilizers applied to them, and dust. Anthropogenic nitrogen inputs account for about 80% of all nitrogen fluxes in ecosystems.

When plant tissues are shed or are eaten, the nitrogen in those tissues becomes available to animals and microbes. Microbial decomposition releases nitrogen compounds from dead organic matter in the soil, where plants, fungi and bacteria compete for it. Some soil bacteria use organic nitrogen-containing compounds as a source of carbon, and release ammonium ions into the soil. This process is known as nitrogen mineralization. Others convert ammonium to nitrite and nitrate ions, a process known as nitrification. Nitric oxide and nitrous oxide are also produced during nitrification. Under nitrogen-rich and oxygen-poor conditions, nitrates and nitrites are converted to nitrogen gas, a process known as denitrification.

Other important nutrients include phosphorus, sulphur, calcium, potassium, magnesium and manganese. Phosphorus enters ecosystems through weathering. As ecosystems age this supply diminishes, making phosphorus-limitation more common in older landscapes (especially in the tropics). Calcium and sulphur are also produced by weathering, but acid deposition is an important source of sulphur in many ecosystems. Although magnesium and manganese are produced by weathering, exchanges between soil organic matter and living cells account for a significant portion of ecosystem fluxes. Potassium is primarily cycled between living cells and soil organic matter.

Function and Biodiversity

Ecosystem processes are broad generalizations that actually take place through the actions of individual organisms. The nature of the organisms—the species, functional groups and trophic levels to which they belong—dictates the sorts of actions these individuals are capable of carrying out, and the relative efficiency with which they do so. Thus, ecosystem processes are driven by the number of species in an ecosystem, the exact nature of each individual species, and the relative abundance organisms within these species. Biodiversity plays an important role in ecosystem functioning.

Ecological theory suggests that in order to coexist, species must have some level of limiting similarity—they must be different from one another in some fundamental way, otherwise one species would competitively exclude the other. Despite this, the cumulative effect of additional species in an ecosystem is not linear additional species may enhance nitrogen retention, for example, but beyond some level of species richness, additional species may have little additive effect. The addition (or loss) of species which are ecologically similar to those already present in an ecosystem tends to only have a small effect on ecosystem function. Ecologically distinct species, on the other hand, have a much larger effect. Similarly, dominant species have a large impact on ecosystem function, while rare species tend to have a small effect. Keystone species tend to have an effect on ecosystem function that is disproportionate to their abundance in an ecosystem.

Ecosystem Goods and Services

Ecosystems provide a variety of goods and services upon which people depend. Ecosystem goods include the "tangible, material products" of ecosystem processes—food, construction material, medicinal plants—in addition to less tangible items like tourism and recreation, and genes from wild plants and animals that can be used to improve domestic species. Ecosystem services, on the other hand, are generally "improvements in the condition or location of things of value". These include things like the maintenance of hydrological cycles, cleaning air and water, the maintenance of oxygen in the atmosphere, crop pollination and even things like beauty, inspiration and opportunities for research. While ecosystem goods have traditionally been recognized as being the basis for things of economic value, ecosystem services tend to be taken for granted. While Gretchen Daily's original definition distinguished between ecosystem goods and ecosystem services, Robert Costanza and colleagues' later work and that of the Millennium Ecosystem Assessment lumped all of these together as ecosystem services.

Ecosystem Management

When natural resource management is applied to whole ecosystems, rather than single species, it is termed ecosystem management. A variety of definitions exist: F. Stuart Chapin and coauthors define it as "the application of ecological science to resource management to promote long-term sustainability of ecosystems and the delivery of essential ecosystem goods and services", while Norman Christensen and coauthors defined it as "management driven by explicit goals, executed by policies, protocols, and practices, and made adaptable by monitoring and research based on our best understanding of the ecological interactions and processes necessary to sustain ecosystem structure and function" and Peter Brussard and colleagues defined it as "managing areas at various scales in such a way that ecosystem services and biological resources are preserved while appropriate human use and options for livelihood are sustained".

Although definitions of ecosystem management abound, there is a common set of principles which underlie these definitions. A fundamental principle is the long-term sustainability of the production of goods and services by the ecosystem; "intergenerational sustainability [is] a precondition for management, not an afterthought". It also requires clear goals with respect to future trajectories and behaviours of the system being managed. Other important requirements include a sound ecological understanding of the system, including connectedness, ecological dynamics and the context in which the system is embedded. Other important principles include an understanding of the role of humans as components of the ecosystems and the use of adaptive management. While ecosystem management can be used as part of a plan for wilderness conservation, it can also be used in intensively managed ecosystems.

Ecosystem Dynamics

Ecosystems are dynamic entities—invariably, they are subject to periodic disturbances and are in the process of recovering from some past disturbance. When an ecosystem is subject to some sort of perturbation, it responds by moving away from its initial state. The tendency of a system to remain close to its equilibrium state, despite that disturbance, is termed its resistance. On the other hand, the speed with which it returns to its initial state after disturbance is called its resilience.

From one year to another, ecosystems experience variation in their biotic and abiotic environments. A drought, an especially cold winter and a pest outbreak all constitute short-term variability in environmental conditions. Animal populations vary from year to year, building up during resource-rich periods and crashing as they overshoot their food supply. These changes play out in changes in NPP, decomposition rates, and other ecosystem processes. Longer-term changes also shape ecosystem processes—the forests of eastern North America still show legacies of cultivation which ceased 200 years ago, while methane production in eastern Siberian lakes is controlled by organic matter which accumulated during the Pleistocene.

Disturbance also plays an important role in ecological processes. F. Stuart Chapin and coauthors define disturbance as "a relatively discrete event in time and space that alters the structure of populations, communities and ecosystems and causes changes in resource availability or the physical environment". This can range from tree falls and insect outbreaks to hurricanes and wildfires to volcanic eruptions and can cause large changes in plant, animal and microbe populations, as well soil organic matter content. Disturbance is followed by succession, a "directional change in ecosystem structure and functioning resulting from biotically driven changes in resources supply." The frequency and severity of disturbance determines the way it impacts ecosystem function. Major disturbance like a volcanic eruption or glacial advance and retreat leave behind soils that lack plants, animals or organic matter. Ecosystems that experience disturbances that sever undergo primary succession. Less severe disturbance like forest fires, hurricanes or cultivation result in secondary succession. More severe disturbance and more frequent disturbance result in longer recovery times. Ecosystems recover more quickly from less severe disturbance events.

The early stages of primary succession are dominated by species with small propagules which can be dispersed long distances. The early colonizers—often algae, cyanobacteria and lichens—stabilize the substrate. Nitrogen supplies are limited in new soils, and nitrogen-fixing species tend to play an important role early in primary succession. Unlike in primary succession, the species that dominate secondary succession, are usually present from the start of the process, often in the soil seed bank. In some systems the successional pathways are fairly consistent, and thus, are easy to predict. In others, there are many possible pathways—for example, the introduced nitrogen-fixing legume, *Myrica faya*, alter successional trajectories in Hawai'ian forests.

The theoretical ecologist Robert Ulanowicz has used information theory tools to describe the structure of ecosystems, emphasizing mutual information (correlations) in studied systems. Drawing on this methodology and prior observations of complex ecosystems, Ulanowicz depicts approaches to determining the stress levels on ecosystems and predicting system reactions to defined types of alteration in their settings (such as increased or reduced energy flow, and eutrophication.

Ecosystem Ecology

Ecosystem ecology studies "the flow of energy and materials through organisms and the physical environment". It seeks to understand the processes which govern the stocks of material and energy in ecosystems, and the flow of matter and energy through them. The study of ecosystems can cover 10 orders of magnitude, from the surface layers of rocks to the surface of the planet.

There is no single definition of what constitutes an ecosystem. German ecologist Ernst-Detlef Schulze and coauthors defined an ecosystem as an area which is “uniform regarding the biological turnover, and contains all the fluxes above and below the ground area under consideration.” They explicitly reject Gene Likens' use of entire river catchments as “too wide a demarcation” to be a single ecosystem, given the level of heterogeneity within such an area. Other authors have suggested that an ecosystem can encompass a much larger area, even the whole planet. Schulze and coauthors also rejected the idea that a single rotting log could be studied as an ecosystem because the size of the flows between the log and its surroundings are too large, relative to the proportion cycles within the log. Philosopher of science Mark Sagoff considers the failure to define “the kind of object it studies” to be an obstacle to the development of theory in ecosystem ecology.

Ecosystems can be studied through a variety of approaches—theoretical studies, studies monitoring specific ecosystems over long periods of time, those that look at differences between ecosystems to elucidate how they work and direct manipulative experimentation. Studies can be carried out at a variety of scales, from microcosms and mesocosms which serve as simplified representations of ecosystems, through whole-ecosystem studies. American ecologist Stephen R. Carpenter has argued that microcosm experiments can be “irrelevant and diversionary” if they are not carried out in conjunction with field studies carried out at the ecosystem scale, because microcosm experiments often fail to accurately predict ecosystem-level dynamics.

The Hubbard Brook Ecosystem Study, established in the White Mountains, New Hampshire in 1963, was the first successful attempt to study an entire watershed as an ecosystem. The study used stream chemistry as a means of monitoring ecosystem properties, and developed a detailed biogeochemical model of the ecosystem. Long-term research at the site led to the discovery of acid rain in North America in 1972, and was able to document the consequent depletion of soil cations (especially calcium) over the next several decades

Classification

Classifying ecosystems into ecologically homogeneous units is an important step towards effective ecosystem management. A variety of systems exist, based on vegetation cover, remote sensing, and bioclimatic classification systems. American geographer Robert Bailey defines a hierarchy of ecosystem units ranging from microecosystems (individual homogeneous sites, on the order of 10 square kilometres (4 sq mi) in area), through mesoecosystems (landscape mosaics, on the order of 1,000 square kilometres (400 sq mi)) to macroecosystems (ecoregions, on the order of 100,000 square kilometres (40,000 sq mi)).

Bailey outlined five different methods for identifying ecosystems: *gestalt* ("a whole that is not derived through considerable of its parts"), in which regions are recognized and boundaries drawn intuitively; a map overlay system where different layers like geology, landforms and soil types are overlain to identify ecosystems; mulitvariate clustering of site attributes; digital image processing of remotely sensed data grouping areas based on their appearance or other spectral properties; or by a "controlling factors method" where a subset of factors (like soils, climate, vegetation physiognomy or the distribution of plant or animal species) are selected from a large array of possible ones are used to delineate ecosystems.

In contrast with Bailey's methodology, Puerto Rico ecologist Ariel Lugo and coauthors identified ten characteristics of an effective classification system: that it be based on georeferenced, quantitative data; that it should minimize subjectivity and explicitly identify criteria and assumptions; that it should be structured around the factors that drive ecosystem processes; that it should reflect the hierarchical nature of ecosystems; that it should be flexible enough to conform to the various scales at which ecosystem management operates; that it should be tied to reliable measures of climate so that it can "anticipat[e] global climate change; that it be applicable worldwide; that it should be validated against independent data; that it take into account the sometimes complex relationship between climate, vegetation and ecosystem functioning; and that it should be able to adapt and improve as new data become available".

Types of Ecosystems

Aquatic Ecosystem: An aquatic ecosystem is an ecosystem in a body of water. Communities of organisms that are dependent on each other and on their environment live in aquatic ecosystems. The two main types of aquatic ecosystems are marine ecosystems and freshwater ecosystems.

***Marine Ecosystems*:** Marine ecosystems are among the largest of Earth's aquatic ecosystems. They include oceans, salt marsh and intertidal ecology, estuaries and lagoons, mangroves and coral reefs, the deep sea and the sea floor. They can be contrasted with freshwater ecosystems, which have a lower salt content. Marine waters cover two-thirds of the surface of the Earth. Such places are considered ecosystems because the plant life supports the animal life and vice-versa.

Marine ecosystems are very important for the overall health of both marine and terrestrial environments. According to the World Resource Center, coastal habitats alone account for approximately 1/3 of all marine biological productivity, and estuarine ecosystems (i.e., salt marshes, seagrasses, mangrove forests) are among the most productive regions on the planet. In addition, other marine ecosystems such as coral reefs, provide food and shelter to the highest levels of marine diversity in the world.

Marine ecosystems usually have a large biodiversity and are therefore thought to have a good resistance against invasive species. However, exceptions have been observed, and the mechanisms responsible in determining the success of an invasion are not yet clear.

Marine ecosystems cover approximately 71% of the Earth's surface and contain approximately 97% of the planet's water. They generate 32% of the world's net primary production. They are distinguished from freshwater ecosystems by the presence of dissolved compounds, especially salts, in the water. Approximately 85% of the dissolved materials in seawater are sodium and chlorine. Seawater has an average salinity of 35 parts per thousand (ppt) of water. Actual salinity varies among different marine ecosystems.

Marine ecosystems can be divided into many zones depending upon water depth and shoreline features. The oceanic zone is the vast open part of the ocean where animals such as whales, sharks, and tuna live. The benthic zone consists of substrates below water where many invertebrates live. The intertidal zone is the area between high and low tides; in this figure it is termed the littoral zone. Other near-shore (neritic) zones can include estuaries, salt marshes, coral reefs, lagoons and mangrove swamps. In the deep water, hydrothermal vents may occur where chemosynthetic sulphur bacteria form the base of the food web.

Classes of organisms found in marine ecosystems include brown algae, dinoflagellates, corals, cephalopods, echinoderms, and sharks. Fishes caught in marine ecosystems are the biggest source of commercial foods obtained from wild populations. Environmental problems concerning marine ecosystems include unsustainable exploitation of marine resources (for example overfishing of certain species), marine pollution, climate change, and building on coastal areas.

Freshwater Ecosystems: Freshwater ecosystems are a subset of Earth's aquatic ecosystems. They include lakes and ponds, rivers, streams and springs, and wetlands. They can be contrasted with marine ecosystems, which have a larger salt content. Freshwater habitats can be classified by different factors, including temperature, light penetration, and vegetation. Freshwater ecosystems can be divided into lentic ecosystems (still water) and lotic ecosystems (flowing water). Limnology (and its branch freshwater biology) is a study about freshwater ecosystems. It is a part of hydrobiology. Original efforts to understand and monitor freshwater ecosystems were spurred on by threats to human health (ex. Cholera outbreaks due to sewage contamination). Early monitoring focussed on chemical indicators, then bacteria, and finally algae, fungi and protozoa. A new type of monitoring involves differing groups of organisms (macroinvertebrates, macrophytes and fish) and the stream conditions associated with them.

Current biomonitering techniques focus mainly on community structure or biochemical oxygen demand. Responses are measured by behavioural changes, altered rates of growth, reproduction or mortality. Macroinvertebrates are most often used in these models because of well known taxonomy, ease of collection, sensitivity to a range of stressors, and their overall value to the ecosystem. Most of these measurements are difficult to extrapolate on a large scale however.

The use of reference sites is common when assessing what a healthy freshwater ecosystem should "look like". Reference sites are easier to reconstruct in standing water than moving water. Preserved indicators such as diatom valves, macrophyte pollen, insect chitin and fish scales can be used to establish a reference ecosystem representative of a time before large scale human distu-rbance.

Common chemical stresses on freshwater ecosystem health include acidification, eutrophication and copper and pesticide contamination.

Extinction of Freshwater Fauna

Over 123 freshwater fauna species have gone extinct in North America since 1900. Of North American freshwater species, an estimated 48.5% of mussels, 22.8% of gastropods, 32.7% of crayfishes, 25.9% of amphibians, and 21.3% of fishes are either endangered or threatened. Extinction rates of many species may increase severely into the next century because of invasive species, loss of keystone species and species which are already functionally extinct. Projected extinction rates for freshwater animals are around five times greater than for land animals, and are comparable to the rates for rainforest communities. Recent extinction trends can be attributed largely to sedimentation, stream fragmentation, chemical and organic pollutants, dams, and invasive species.

Freshwater ecosystems cover 0.80% of the Earth's surface and inhabit 0.009% of its total water. They generate nearly 3% of its net primary production. Freshwater ecosystems contain 41% of the world's known fish species.

There are three basic types of freshwater ecosystems:

- Lentic: slow moving water, including pools, ponds, and lakes.
- Lotic: faster moving water, for example streams and rivers.
- Wetlands: areas where the soil is saturated or inundated for at least part of the time.

Lentic

Lake ecosystems can be divided into zones. One common system divides lakes into three zones. The first, the littoral zone, is the shallow zone near the shore. This is where rooted wetland plants occur. The offshore is divided into two further zones, an open water zone and a deep water zone. In the open water zone (or photic zone) sunlight supports photosynthetic algae, and the species that feed upon them. In the deep water zone, sunlight is not available and the food web is based on detritus entering from the littoral and photic zones. Some systems use other names.

The off shore areas may be called the pelagic zone, and the aphotic zone may be called the profundal zone. Inland from the littoral zone one can also frequently identify a riparian zone which has plants still affected by the presence of the lake—this can include effects from windfalls, spring flooding, and winter ice damage. The production of the lake as a whole is the result of production from plants growing in the littoral zone, combined with production from plankton growing in the open water.

Wetlands can be part of the lentic system, as they form naturally along most lakeshores, the width of the wetland and littoral zone being dependent upon the slope of the shoreline and the amount of natural change in water levels, within and among years. Often dead trees accumulate in this zone, either from windfalls on the shore or logs transported to the site during floods. This woody debris provides important habitat for fish and nesting birds, as well as protecting shorelines from erosion,

Two important subclasses of lakes are ponds, which typically are small lakes that intergrade with wetlands, and water reservoirs. Over long periods of time, lakes, or bays within them, may gradually become enriched by nutrients and slowly fill in with organic sediments, a process called succession. When humans use the watershed, the volumes of sediment entering the lake can accelerate this process. The addition of sediments and nutrients to a lake is known as eutrophication.

Ponds

Ponds are small bodies of freshwater with shallow and still water, marsh, and aquatic plants. They can be further divided into four zones: vegetation zone, open water, bottom mud and surface film. The size and depth of ponds often varies greatly with the time of year; many ponds are produced by spring flooding from rivers. Food webs are based both on free-floating algae and upon aquatic plants. There is usually a diverse array of aquatic life, with a few examples including algae, snails, fish, beetles, water bugs, frogs, turtles, otters and muskrats. Top predators may include large fish, herons, or alligators. Since fish are a major predator upon amphibian larvae, ponds that dry up each year, thereby killing resident fish, provide important refugia for amphibian breeding. Ponds that dry up completely each year are often known as vernal pools. Some ponds are produced by animal activity, including alligator holes and beaver ponds, and these add important diversity to landscapes.

River Ecosystem

The ecosystem of a river is the river viewed as a system operating in its natural environment, and includes biotic (living) interactions amongst plants, animals and micro-organisms, as well as abiotic (nonliving) physical and chemical interactions.

River ecosystems are prime examples of lotic ecosystems. *Lotic* refers to flowing water, from the Latin *lotus*, washed. Lotic waters range from springs only a few centimeters wide to major rivers kilometers in width. Much of this article applies to lotic ecosystems in general, including related lotic systems such as streams and springs. Lotic ecosystems can be contrasted with lentic ecosystems, which involve relatively still terrestrial waters such as lakes and ponds. Together, these two fields form the more general study area of freshwater or aquatic ecology.

The major zones in river ecosystems are determined by the river bed's gradient or by the velocity of the current. Faster moving turbulent water typically contains greater concentrations of dissolved oxygen, which supports greater biodiversity than the slow moving water of pools. These distinctions forms the basis for the division of rivers into upland and lowland rivers. The food base of streams within riparian forests is mostly derived from the trees, but wider streams and those that lack a canopy derive the majority of their food base from algae. Anadromous fish are also an important source of nutrients. Environmental threats to rivers include loss of water, dams, chemical pollution and introduced species. A dam produces negative effects that continue down the watershed. The most important negative effects are the reduction of spring flooding, which damages wetlands, and the retention of sediment, which leads to loss of deltaic wetlands.

The following unifying characteristics make the ecology of running waters unique from that of other aquatic habitats.

- Flow is unidirectional.
- There is a state of continuous physical change.
- There is a high degree of spatial and temporal heterogeneity at all scales (microhabitats).
- Variability between lotic systems is quite high.
- The biota is specialized to live with flow conditions.

Abiotic Factors

Flow: Water flow is the key factor in lotic systems influencing their ecology. The strength of water flow can vary between systems, ranging from torrential rapids to slow backwaters that almost seem like lentic systems. The speed of the water flow can also vary within a system and is subject to chaotic turbulence. This turbulence results in divergences of flow from the mean downslope flow vector as typified by eddy currents. The mean flow rate vector is based on variability of friction with the bottom or sides of the channel, sinuosity, obstructions, and the incline gradient. In addition, the amount of water input into the system from direct precipitation, snowmelt, and/or groundwater can affect flow rate. Flowing waters can alter the shape of the streambed through erosion and deposition, creating a variety of habitats, including riffles, glides, and pools.

Light: Light is important to lotic systems, because it provides the energy necessary to drive primary production via photosynthesis, and can also provide refuge for prey species in shadows it casts. The amount of light that a system receives can be related to a combination of internal and external stream variables. The area surrounding a small stream, for example, might be shaded by surrounding forests or by valley walls. Larger river systems tend to be wide so the influence of external variables is minimized, and the sun reaches the surface. These rivers also tend to be more turbulent, however, and particles in the water increasingly attenuate light as depth increases. Seasonal and diurnal factors might also play a role in light availability because the angle of incidence, the angle at which light strikes water can lead to light lost from reflection. Known as Beer's Law, the shallower the angle, the more light is reflected and the amount of solar radiation received declines logarithmically with depth. Additional influences on light availability include cloud cover, altitude, and geographic position (Brown 1987).

Temperature: Most lotic species are poikilotherms whose internal temperature varies with their environment, thus temperature is a key abiotic factor for them. Water can be heated or cooled through radiation at the surface and conduction to or from the air and surrounding substrate. Shallow streams are typically well mixed and maintain a relatively uniform temperature within an area. In deeper, slower moving water systems, however, a strong difference between the bottom and surface temperatures may develop. Spring fed systems have little variation as springs are typically from groundwater sources, which are often very close to ambient temperature. Many systems show strong diurnal fluctuations and seasonal variations are most extreme in arctic, desert and temperate systems. The amount of shading, climate and elevation can also influence the temperature of lotic systems.

Chemistry: Water chemistry between systems varies tremendously. The chemistry is foremost determined by inputs from the geology of its watershed, or catchment area, but can also be influenced by precipitation and the addition of pollutants from human sources. Large differences in chemistry do not usually exist within small lotic systems due to a high rate of mixing. In larger river systems, however, the concentrations of most nutrients, dissolved salts, and pH decrease as distance increases from the river's source.

Oxygen is likely the most important chemical constituent of lotic systems, as all aerobic organisms require it for survival. It enters the water mostly via diffusion at the water-air interface. Oxygen's solubility in water decreases as water temperature increases. Fast, turbulent streams expose more of the water's surface area to the air and tend to have low temperatures and thus more oxygen than slow, backwaters. Oxygen is a byproduct of photosynthesis, so systems with a high abundance of aquatic algae and plants may also have high concentrations of oxygen during the day. These levels can decrease significantly during the night when primary producers switch to respiration. Oxygen can be limiting if circulation between the surface and deeper layers is poor, if the activity of lotic animals is very high, or if there is a large amount of organic decay occurring.

Substrate

The inorganic substrate of lotic systems is composed of the geologic material present in the catchment that is eroded, transported, sorted, and deposited by the current. Inorganic substrates are classified by size on the Wentworth scale, which ranges from boulders, to pebbles, to gravel, to sand, and to silt. Typically, particle size decreases downstream with larger boulders and stones in more mountainous areas and sandy bottoms in lowland rivers. This is because the higher gradients of mountain streams facilitate a faster flow, moving smaller substrate materials further downstream for deposition. Substrate can also be organic and may include fine particles, autumn shed leaves, submerged wood, moss, and more evolved plants. Substrate deposition is not necessarily a permanent event, as it can be subject to large modifications during flooding events.

Biota

Bacteria: Bacteria are present in large numbers in lotic waters. Free-living forms are associated with decomposing organic material, biofilm on the surfaces of rocks and vegetation, in between particles that compose the substrate, and suspended in the water column. Other forms are also associated with the guts of lotic organisms as parasites or in commensal relationships. Bacteria play a large role in energy recycling, which will be discussed in the Trophic Relationships section

Primary Producers

Algae, consisting of phytoplankton and periphyton, are the most significant sources of primary production in most streams and rivers. Phytoplankton float freely in the water column and thus are unable to maintain populations in fast flowing streams. They can, however, develop sizable populations in slow moving rivers and backwaters. Periphyton are typically filamentous and tufted algae that can attach themselves to objects to avoid being washed away by fast current. In places where flow rates are negligible or absent, periphyton may form a gelatinous, unanchored floating mat. Plants exhibit limited adaptations to fast flow and are most successful in reduced currents. More primitive plants, such as mosses and liverworts attach themselves to solid objects. This typically occurs in colder headwaters where the mostly rocky substrate offers attachment sites. Some plants are free floating at the water's surface in dense mats like duckweed or water hyacinth. Others are rooted and may be classified as submerged or emergent. Rooted plants usually occur in areas of slackened current where fine-grained soils are found (Brown 1987). These rooted plants are flexible, with elongated leaves that offer minimal resistance to current.

Living in flowing water can be beneficial to plants and algae because the current is usually well aerated and it provides a continuous supply of nutrients. These organisms are limited by flow, light, water chemistry, substrate, and grazing pressure. Algae and plants are important to lotic systems as sources of energy, for forming microhabitats that shelter other fauna from predators and the current, and as a food resource (Brown 1987).

Insects and other Invertebrates

Up to 90% of invertebrates in some lotic systems are insects. These species exhibit tremendous diversity and can be found occupying almost every available habitat, including the surfaces of stones, deep below the substratum, adrift in the current, and in the surface film. Insects have developed several strategies for living in the diverse flows of lotic systems. Some avoid high current areas, inhabiting the substratum or the sheltered side of rocks. Additional invertebrate taxa common to flowing waters include mollusks such as snails, limpets, clams, mussels, as well as crustaceans like crayfish and crabs. Like most of the primary consumers, lotic invertebrates often rely heavily on the current to bring them food and oxygen (Brown 1987). Invertebrates, especially insects, are important as both consumers and prey items in lotic systems.

Fish and other Vertebrates

Fishes are probably the best-known inhabitants of lotic systems. The ability of a fish species to live in flowing waters depends upon the speed at which it can swim and the duration that its speed can be maintained. This ability can vary greatly between species and is tied to the habitat in which it can survive. Continuous swimming expends a tremendous amount of energy and, therefore, fishes spend only short periods in full current. Instead, individuals remain close to the bottom or the banks, behind obstacles, and sheltered from the current, swimming in the current only to feed or change locations. Some species have adapted to living only on the system bottom, never venturing into the open water flow. These fishes are dorso-ventrally flattened to reduce flow resistance and often have eyes on top of their heads to observe what is happening above them. Some also have sensory barrels positioned under the head to assist in the testing of substratum (Brown 1987).

Lotic systems typically connect to each other, forming a path to the ocean (spring '! stream '! river '! ocean), and many fishes have life cycles that require stages in both fresh and salt water. Salmon, for example, are anadromous species that are born and develop in freshwater and then move to the ocean as adults. Eels are catadromous, and are born and develop in the ocean and then move into freshwater as adults. Other vertebrate taxa that inhabit lotic systems include amphibians, such as salamanders, reptiles (e.g. snakes, turtles, crocodiles and alligators) various bird species, and mammals (e.g., otters, beavers, hippos, and river dolphins). With the exception of a few species, these vertebrates are not tied to water as fishes are, and spend part of their time in terrestrial habitats. Many fish species are important as consumers and as prey species to the larger vertebrates mentioned above.

Trophic Relationships

Energy Inputs: Energy sources can be autochthonous or allochthonous.

- Autochthonous energy sources are those derived from within the lotic system. During photosynthesis, for example, primary producers form organic carbon compounds out of carbon dioxide and inorganic matter. The energy they produce is important for the community because it may be transferred to higher trophic levels via consumption. Additionally, high rates of primary production can introduce dissolved organic matter (DOM) to the waters. Another form of autochthonous energy comes from the decomposition of dead organisms and feces that originate within the lotic system. In this case, bacteria decompose the detritus or coarse particulate organic material (CPOM, >1 mm pieces) into fine particulate organic matter (FPOM; <1 mm pieces) and then further into inorganic compounds that are required for photosynthesis. This process is discussed in more detail below.

- Allochthonous energy sources are those derived from outside the lotic system, that is, from the terrestrial environment. Leaves, twigs, fruits, etc. are typical forms of terrestrial CPOM that have entered the water by direct litterfall or lateral leaf blow. In addition, terrestrial animal-derived materials, such as feces or carcasses that have been added to the system are examples of allochthonous CPOM. The CPOM undergoes a specific process of degradation. Allan gives the example of a leaf fallen into a stream. First, the soluble chemicals are dissolved and leached from the leaf upon its saturation with water. This adds to the DOM load in the system. Next, microbes such as bacteria and fungi colonize the leaf, softening it as the mycelium of the fungus grows into it. The composition of the microbial community is influenced by the species of tree from which the leaves are shed (Rubbo and Kiesecker 2004). This combination of bacteria, fungi, and leaf are a food source for shredding invertebrates, which leave only FPOM after consumption. These fine particles may be colonized by microbes again or serve as a food source for animals that consume FPOM. Organic matter can also enter the lotic system already in the FPOM stage by wind, surface runoff, bank erosion, or groundwater. Similarly, DOM can be introduced through canopy drip from rain or from surface flows.

Invertebrates

Invertebrates can be organized into many feeding guilds in lotic systems. Some species are shredders, which use large and powerful mouth parts to feed on non-woody CPOM and their associated microorganisms. Others are suspension feeders, which use their setae, filtering aparati, nets, or even secretions to collect FPOM and microbes from the water. These species may be passive collectors, utilizing the natural flow of the system, or they may generate their own current to draw water, and also, FPOM in Allan. Members of the gatherer-collector guild actively search for FPOM under rocks and in other places where the stream flow has slackened enough to allow deposition. Grazing invertebrates utilize scraping, rasping, and browsing adaptations to feed on periphyton and detritus. Finally, several families are predatory, capturing and consuming animal prey. Both the number of species and the abundance of individuals within each guild is largely dependent upon food availability. Thus, these values may vary across both seasons and systems.

Fish

Fish can also be placed into feeding guilds. Planktivores pick plankton out of the water column. Herbivore-detritivores are bottom-feeding species that ingest both periphyton and detritus indiscriminately. Surface and water column feeders capture surface prey (mainly terrestrial and emerging insects) and drift (benthic invertebrates floating downstream). Benthic invertebrate feeders prey primarily on immature insects, but will also consume other benthic invertebrates. Top predators consume fishes and/or large invertebrates. Omnivores ingest a wide range of prey. These can be floral, faunal, and/or detrital in nature. Finally, parasites live off of host species, typically other fishes. Fish are flexible in their feeding roles, capturing different prey with regard to seasonal availability and their own developmental stage. Thus, they may occupy multiple feeding guilds in their lifetime. The number of species in each guild can vary greatly between systems, with temperate warm water streams having the most benthic invertebrate feeders, and tropical systems having large numbers of detritus feeders due to high rates of allochthonous input.

Community Patterns and Diversity

Local Species Richness: Large rivers have comparatively more species than small streams. Many relate this pattern to the greater area and volume of larger systems, as well as an increase in habitat diversity. Some systems, however, show a poor fit between system size and species richness. In these cases, a combination of factors such as historical rates of speciation and extinction, type of substrate, microhabitat availability, water chemistry, temperature, and disturbance such as flooding seem to be important.

Resource Partitioning

Although many alternate theories have been postulated for the ability of guild-mates to coexist, resource partitioning has been well documented in lotic systems as a means of reducing competition. The three main types of resource partitioning include habitat, dietary, and temporal segregation.

Habitat segregation was found to be the most common type of resource partitioning in natural systems (Schoener, 1974). In lotic systems, microhabitats provide a level of physical complexity that can support a diverse array of organisms (Vincin and Hawknis, 1998). The separation of species by substrate preferences has been well documented for invertebrates. Ward (1992) was able to divide substrate dwellers into 6 broad assemblages, including those that live in: coarse substrate, gravel, sand, mud, woody debris, and those associated with plants, showing one layer of segregation. On a smaller scale, further habitat partitioning can occur on or around a single substrate, such as a piece of gravel. Some invertebrates prefer the high flow areas on the exposed top of the gravel, while others reside in the crevices between one piece of gravel and the next, while still others live on the bottom of this gravel piece.

Dietary segregation is the second-most common type of resource partitioning. High degrees of morphological specializations or behavioural differences allow organisms to use specific resources. The size of nets built by some species of invertebrate suspension feeders, for example, can filter varying particle size of FPOM from the water (Edington et al. 1984). Similarly, members in the grazing guild can specialize in the harvesting of algae or detritus depending upon the morphology of their scraping apparatus. In addition, certain species seem to show a preference for specific algal species.

Temporal segregation is a less common form of resource partitioning, but it is nonetheless an observed phenomenon. Typically, it accounts for coexistence by relating it to differences in life history patterns and the timing of maximum growth among guild mates. Tropical fishes in Borneo, for example, have shifted to shorter life spans in response to the ecological niche reduction felt with increasing levels of species richness in their ecosystem (Watson and Balon 1984).

Persistence and Succession

Over long time scales, there is a tendency for species composition in pristine systems to remain in a stable state. This has been found for both invertebrate and fish species. On shorter times scales, however, flow variability and unusual precipitation patterns decrease habitat stability and can all lead to declines in persistence levels. The ability to maintain this persistence over long time scales is related to the ability of lotic systems to return to the original community configuration relatively quickly after a disturbance (Townsend et al. 1987). This is one example of temporal succession, a site-specific change in a community involving changes in species composition over time. Another form of temporal succession might occur when a new habitat is opened up for colonization. In these cases, an entirely new community that is well adapted to the conditions found in this new area can establish itself.

River Continuum Concept

The River continuum concept (RCC) was an attempt to construct a single framework to describe the function of temperate lotic ecosystems from the source to the end and relate it to changes in the biotic community (Vannote et al. 1980). The physical basis for RCC is size and location along the gradient from a small stream eventually linked to a large river. Stream order is used as the physical measure of the position along the RCC.

According to the RCC, low ordered sites are small shaded streams where allochthonous inputs of CPOM are a necessary resource for consumers. As the river widens at mid-ordered sites, energy inputs should change. Ample sunlight should reach the bottom in these systems to support significant periphyton production. Additionally, the biological processing of CPOM (Coarse Particulate Organic Matter - larger than 1 mm) inputs at upstream sites is expected to result in the transport of large amounts of FPOM (Fine Particulate Organic Matter - smaller than 1 mm) to these downstream ecosystems. Plants should become more abundant at edges of the river with increasing river size, especially in lowland rivers where finer sediments have been deposited and facilitate rooting. The main channels likely have too much current and turbidity and a lack of substrate to support plants or periphyton. Phytoplankton should produce the only autochthonous inputs here, but photosynthetic rates will be limited due to turbidity and mixing. Thus, allochthonous inputs are expected to be the primary energy source for large rivers. This FPOM will come from both upstream sites via the decomposition process and through lateral inputs from floodplains.

Biota should change with this change in energy from the headwaters to the mouth of these systems. Namely, shredders should prosper in low-ordered systems and grazers in mid-ordered sites. Microbial decomposition should play the largest role in energy production for low-ordered sites and large rivers, while photosynthesis, in addition to degraded allochthonous inputs from upstream will be essential in mid-ordered systems. As mid-ordered sites will theoretically receive the largest variety of energy inputs, they might be expected to host the most biological diversity (Vannote et al. 1980).

Just how well the RCC actually reflects patterns in natural systems is uncertain and its generality can be a handicap when applied to diverse and specific situations. The most noted criticisms of the RCC are: 1. It focuses mostly on macroinvertebrates, disregarding that plankton and fish diversity is highest in high orders; 2. It relies heavily on the fact that low ordered sites have high CPOM inputs, even though many streams lack riparian habitats; 3. It is based on pristine systems, which rarely exist today; and 4. It is centered around the functioning of temperate streams. Despite its shortcomings, the RCC remains a useful idea for describing how the patterns of ecological functions in a lotic system can vary from the source to the mouth.

Disturbances such as congestion by dams or natural events such as shore flooding are not included in the RCC model. Various researchers have since expanded the model to account for such irregularities. For example, J.V. Ward and J.A. Stanford came up with the Serial Discontinuity Concept in 1983, which addresses the impact of geomorphologic disorders such as congestion and integrated inflows. The same authors presented the Hyporheic Corridor concept in 1993, in which the vertical (in depth) and lateral (from shore to shore) structural complexity of the river were connected. The flood pulse concept, developed by W.J. Junk in 1989, further modified by P.B. Bayley in 1990 and K. Tockner in 2000, takes into account the large amount of nutrients and organic material that makes its way into a river from the sediment of surrounding flooded land.

Human Impacts

Pollution: Pollutant sources of lotic systems are hard to control because they derive, often in small amounts, over a very wide area and enter the system at many locations along its length. Agricultural fields often deliver large quantities of sediments, nutrients, and chemicals to nearby streams and rivers. Urban and residential areas can also add to this pollution when contaminants are accumulated on impervious surfaces such as roads and parking lots that then drain into the system. Elevated nutrient concentrations, especially nitrogen and phosphorus which are key components of fertilizers, can increase periphyton growth, which can be particularly dangerous in slow moving streams. Another pollutant, acid rain, forms from sulphur dioxide and nitrous oxide emitted from factories and power stations. These substances readily dissolve in atmospheric moisture and enter lotic systems through precipitation. This can lower the pH of these sites, affecting all trophic levels from algae to vertebrates (Brown 1987). Mean species richness and total species numbers within a system decrease with decreasing pH.

While direct pollution of lotic systems has been greatly reduced in the United States under the government's Clean Water Act, contaminants from diffuse, non-point sources, remain a large problem.

Flow Modification

Dams alter the flow, temperature, and sediment regime of lotic systems. Additionally, many rivers are dammed at multiple locations, amplifying the impact. Dams can cause enhanced clarity and reduced variability in stream flow, which in turn cause an increase in periphyton abundance. Invertebrates immediately below a dam can show reductions in species richness due to an overall reduction in habitat heterogeneity. Also, thermal changes can affect insect development, with abnormally warm winter temperatures obscuring cues to break egg diapause and overly cool summer temperatures leaving too few acceptable days to complete growth. Finally, dams fragment river systems, isolating previously continuous populations, and preventing the migrations of anadromous and catadromous species.

Invasive Species

Invasive species have been introduced to lotic systems through both purposeful events (e.g. stocking game and food species) as well as unintentional events (e.g. hitchhikers on boats or fishing waders). These organisms can affect natives via competition for prey or habitat, predation, habitat alteration, hybridization, or the introduction of harmful diseases and parasites. Once established, these species can be difficult to control or eradicate, particularly because of the connectivity of lotic systems. Invasive species can be especially harmful in areas that have endangered biota, such as mussels in the Southeast United States, or those that have localized endemic species, like lotic systems west of the Rocky Mountains, where many species evolved in isolation.

Terrestrial Ecosystem

A terrestrial ecosystem is an ecosystem found only on landforms. Six primary terrestrial ecosystems exist: tundra, taiga, temperate deciduous forest, tropical rain forest, grassland and desert. A community of organisms and their environment that occurs on the land masses of continents and islands. Terrestrial ecosystems are distinguished from aquatic ecosystems by the lower availability of water and the consequent importance of water as a limiting factor. Terrestrial ecosystems are characterized by greater temperature fluctuations on both a diurnal and seasonal basis than occur in aquatic ecosystems in similar climates. The availability of light is greater in terrestrial ecosystems than in aquatic ecosystems because the atmosphere is more transparent than water. Gases are more available in terrestrial ecosystems than in aquatic ecosystems. Those gases include carbon dioxide that serves as a substrate for photosynthesis, oxygen that serves as a substrate in aerobic respiration, and nitrogen that serves as a substrate for nitrogen fixation. Terrestrial environments are segmented into a subterranean portion from which most water and ions are obtained, and an atmospheric portion from which gases are obtained and where the physical energy of light is transformed into the organic energy of carbon-carbon bonds through the process of photosynthesis.

Terrestrial ecosystems occupy 55,660,000 mi2 (144,150,000 km2), or 28.2%, of Earth's surface. Although they are comparatively recent in the history of life (the first terrestrial organisms appeared in the Silurian Period, about 425 million years ago) and occupy a much smaller portion of Earth's surface than marine ecosystems, terrestrial ecosystems have been a major site of adaptive radiation of both plants and animals. Major plant taxa in terrestrial ecosystems are members of the division Magnoliophyta (flowering plants), of which there are about 275,000 species, and the division Pinophyta (conifers), of which there are about 500 species. Members of the division Bryophyta (mosses and liverworts), of which there are about 24,000 species, are also important in some terrestrial ecosystems. Major animal taxa in terrestrial ecosystems include the classes Insecta (insects) with about 900,000 species, Aves (birds) with 8500 species, and Mammalia (mammals) with approximately 4100 species.

Organisms in terrestrial ecosystems have adaptations that allow them to obtain water when the entire body is no longer bathed in that fluid, means of transporting the water from limited sites of acquisition to the rest of the body, and means of preventing the evaporation of water from body surfaces. They also have traits that provide body support in the atmosphere, a much less buoyant medium than water, and other traits that render them capable of withstanding the extremes of temperature, wind, and humidity that characterize terrestrial ecosystems. Finally, the organisms in terrestrial ecosystems have evolved many methods of transporting gametes in environments where fluid flow is much less effective as a transport medium. The organisms in terrestrial ecosystems are integrated into a functional unit by specific, dynamic relationships due to the coupled processes of energy and chemical flow. Those relationships can be summarized by schematic diagrams of trophic webs, which place organisms according to their feeding relationships. The base of the food web is occupied by green plants, which are the only organisms capable of utilizing the energy of the Sun and inorganic nutrients obtained from the soil to produce organic molecules. Terrestrial food webs can be broken into two segments based on the status of the plant material that enters them. Grazing food webs are associated with the consumption of living plant material by herbivores. Detritus food webs are associated with the consumption of dead plant material by detritivores. The relative importance of those two types of food webs varies considerably in different types of terrestrial ecosystems. Grazing food webs are more important in grasslands, where over half of net primary productivity may be consumed by herbivores. Detritus food webs are more important in forests, where less than 5% of net primary productivity may be consumed by herbivores.

There is one type of extensive terrestrial ecosystem due solely to human activities and eight types that are natural ecosystems. Those natural ecosystems reflect the variation of precipitation and temperature over Earth's surface. The smallest land areas are occupied by tundra and temperate grassland ecosystems, and the largest land area is occupied by tropical forest. The most productive ecosystems are temperate and tropical forests, and the least productive are deserts and tundras. Cultivated lands, which together with grasslands and savannas utilized for grazing are referred to as agroecosystems, are of intermediate extent and productivity. Because of both their areal extent and their high average productivity, tropical forests are the most productive of all terrestrial ecosystems, contributing 45% of total estimated net primary productivity on land.Only animals that is capable of living in the landforms live here.

Forest

A forest, also referred to as a wood or the woods, is an area with a high density of trees. As with cities, depending on various cultural definitions, what is considered a forest may vary significantly in size and have different classifications according to how and of what the forest is composed. A forest is usually an area filled with trees but any tall densely packed area of vegetation may be considered a forest, even underwater vegetation such as kelp forests, or non-vegetation such as fungi, and bacteria. Tree forests cover approximately 9.4 percent of the Earth's surface (or 30 percent of total land area), though they once covered much more (about 50 percent of total land area). They function as habitats for organisms, hydrologic flow modulators, and soil conservers, constituting one of the most important aspects of the biosphere.

A typical tree forest is composed of the overstory (canopy or upper tree layer) and the understory. The understory is further subdivided into the shrub layer, herb layer, and also the moss layer and soil microbes. In some complex forests, there is also a well-defined lower tree layer. Forests are central to all human life because they provide a diverse range of resources: they store carbon, aid in regulating the planetary climate, purify water and mitigate natural hazards such as floods. Forests also contain roughly 90 percent of the world's terrestrial biodiversity.

The word "forest" comes from Middle English *forest*, from Old French *forest* (also *forès*) "forest, vast expanse covered by trees"; first introduced in English as the word for wild land set aside for hunting without the necessity in definition for the existence of trees (James 1981;Muir 2000,2008). Possibly a borrowing (probably via Frankish or Old High German) of the Medieval Latin word *foresta* "open wood", *foresta* was first used by Carolingian scribes in the Capitularies of Charlemagne to refer specifically to the king's royal hunting grounds. The term was not endemic to Romance languages (e.g. native words for "forest" in the Romance languages evolved out of the Latin word *silva* "forest, wood" (English *sylvan*); cf. Italian, Spanish, Portuguese *selva*; Romanian *silvă*; Old French *selve*); and cognates in Romance languages, such as Italian *foresta*, Spanish and Portuguese *floresta*, etc. are all ultimately borrowings of the French word.

The exact origin of Medieval Latin *foresta* is obscure. Some authorities claim the word derives from the Late Latin phrase *forestam silvam*, meaning "the outer wood"; others claim the term is a latinisation of the Frankish word **forhist* "forest, wooded country", assimilated to *forestam silvam* (a common practice among Frankish scribes). Frankish **forhist* is attested by Old High German *forst* "forest", Middle Low German *vorst* "forest", Old English *fyrhþ* "forest, woodland, game preserve, hunting ground" (English *frith*), and Old Norse *fýri* "coniferous forest", all of which derive from Proto-Germanic **furhísa-*, **furhíþija-* "a fir-wood, coniferous forest", from Proto-Indo-European **perkʷu-* "a coniferous or mountain forest, wooded height".

Uses of the word "forest" in English to denote any uninhabited area of non-enclosure are now considered archaic. The word was introduced by the Norman rulers of England as a legal term (appearing in Latin texts like the Magna Carta) denoting an uncultivated area legally set aside for hunting by feudal nobility. These hunting forests were not necessarily wooded much, if at all. However, as hunting forests did often include considerable areas of woodland, the word "forest" eventually came to mean wooded land more generally. By the start of the fourteenth century the word appeared in English texts, indicating all three senses: the most common one, the legal term and the archaic usage.

Other terms used to mean "an area with a high density of trees" are *wood, woodland, wold, weald, holt, frith* and *firth*. Unlike *forest*, these are all derived from Old English and were not borrowed from another language. Some classifications now reserve the term *woodland* for an area with more open space between trees and distinguish among woodlands, *open forests*, and *closed forests* based on crown cover.

Distribution

Forests can be found in all regions capable of sustaining tree growth, at altitudes up to the tree line, except where natural fire frequency or other disturbance is too high, or where the environment has been altered by human activity.

The latitudes 10° north and south of the Equator are mostly covered in tropical rainforest, and the latitudes between 53°N and 67°N have boreal forest. As a general rule, forests dominated by angiosperms (*broadleaf forests*) are more species-rich than those dominated by gymnosperms (*conifer*, *montane*, or *needleleaf forests*), although exceptions exist.

Forests sometimes contain many tree species only within a small area (as in tropical rain and temperate deciduous forests), or relatively few species over large areas (e.g., taiga and arid montane coniferous forests). Forests are often home to many animal and plant species, and biomass per unit area is high compared to other vegetation communities. Much of this biomass occurs below ground in the root systems and as partially decomposed plant detritus. The woody component of a forest contains lignin, which is relatively slow to decompose compared with other organic materials such as cellulose or carbohydrate.

Forests are differentiated from woodlands by the extent of canopy coverage: in a forest, the branches and the foliage of separate trees often meet or interlock, although there can be gaps of varying sizes within an area referred to as forest. A woodland has a more continuously open canopy, with trees spaced farther apart, which allows more sunlight to penetrate to the ground between them

Among the major forested biomes are:

- rain forest (tropical and temperate)
- taiga
- temperate hardwood forest
- tropical dry forest

Classification

Forests can be classified in different ways and to different degrees of specificity. One such way is in terms of the "biome" in which they exist, combined with leaf longevity of the dominant species (whether they are evergreen or deciduous). Another distinction is whether the forests are composed predominantly of broadleaf trees, coniferous (needle-leaved) trees, or mixed.

- Boreal forests occupy the subarctic zone and are generally evergreen and coniferous.
- Temperate zones support both broadleaf deciduous forests (*e.g.*, temperate deciduous forest) and evergreen coniferous forests (*e.g.*, temperate coniferous forests and temperate rainforests). Warm temperate zones support broadleaf evergreen forests, including laurel forests.
- Tropical and subtropical forests include tropical and subtropical moist forests, tropical and subtropical dry forests, and tropical and subtropical coniferous forests.
- Physiognomy classifies forests based on their overall physical structure or developmental stage (e.g. old growth vs. second growth).
- Forests can also be classified more specifically based on the climate and the dominant tree species present, resulting in numerous different forest types (e.g., ponderosa pine/Douglas-fir forest).

A number of global forest classification systems have been proposed, but none has gained universal acceptance. UNEP-WCMC's forest category classification system is a simplification of other more complex systems (e.g. UNESCO's forest and woodland 'subformations'). This system divides the world's forests into 26 major types, which reflect climatic zones as well as the principal types of trees. These 26 major types can be reclassified into 6 broader categories: temperate needleleaf; temperate broadleaf and mixed; tropical moist; tropical dry; sparse trees and parkland; and forest plantations. Each category is described as a separate section below.

Temperate Needleleaf

Temperate needleleaf forests mostly occupy the higher latitude regions of the northern hemisphere, as well as high altitude zones and some warm temperate areas, especially on nutrient-poor or otherwise unfavourable soils. These forests are composed entirely, or nearly so, of coniferous species (Coniferophyta). In the Northern Hemisphere pines *Pinus*, spruces *Picea*, larches *Larix*, firs *Abies*, Douglas firs

Pseudotsuga and hemlocks *Tsuga*, make up the canopy, but other taxa are also important. In the Southern Hemisphere, most coniferous trees (members of the Araucariaceae and Podocarpaceae) occur in mixtures with broadleaf species, and are classed as broadleaf and mixed forests.

Temperate Broadleaf and Mixed

Temperate broadleaf and mixed forests include a substantial component of trees in the Anthophyta. They are generally characteristic of the warmer temperate latitudes, but extend to cool temperate ones, particularly in the southern hemisphere. They include such forest types as the mixed deciduous forests of the United States and their counterparts in China and Japan, the broadleaf evergreen rainforests of Japan, Chile and Tasmania, the sclerophyllous forests of Australia, central Chile, the Mediterranean and California, and the southern beech Nothofagus forests of Chile and New Zealand.

Tropical Moist

There are many different types of tropical moist forests, although most extensive are the lowland evergreen broad leaf rainforests, for example várzea and igapó forests and the terra firma forests of the Amazon Basin; the peat swamp forests, dipterocarp forests of Southeast Asia; and the high forests of the Congo Basin. Forests located on mountains are also included in this category, divided largely into upper and lower montane formations on the basis of the variation of physiognomy corresponding to changes in altitude.

Tropical Dry

Tropical dry forests are characteristic of areas in the tropics affected by seasonal drought. The seasonality of rainfall is usually reflected in the deciduousness of the forest canopy, with most trees being leafless for several months of the year. However, under some conditions, e.g. less fertile soils or less predictable drought regimes, the proportion of evergreen species increases and the forests are characterised as "sclerophyllous". Thorn forest, a dense forest of low stature with a high frequency of thorny or spiny species, is found where drought is prolonged, and especially where grazing animals are plentiful. On very poor soils, and especially where fire is a recurrent phenomenon, woody savannas develop

Sparse Trees and Parkland

Sparse trees and parkland are forests with open canopies of 10–30% crown cover. They occur principally in areas of transition from forested to non-forested landscapes. The two major zones in which these ecosystems occur are in the boreal region and in the seasonally dry tropics. At high latitudes, north of the main zone of boreal forest or taiga, growing conditions are not adequate to maintain a continuous closed forest cover, so tree cover is both sparse and discontinuous. This vegetation is variously called open taiga, open lichen woodland, and forest tundra. It is species-poor, has high bryophyte cover, and is frequently affected by fire.

Forest Plantations

Forest plantations, generally intended for the production of timber and pulpwood increase the total area of forest worldwide. Commonly mono-specific and/or composed of introduced tree species, these ecosystems are not generally important as habitat for native biodiversity. However, they can be managed in ways that enhance their biodiversity protection functions and they are important providers of ecosystem services such as maintaining nutrient capital, protecting watersheds and soil structure as well as storing carbon. They may also play an important role in alleviating pressure on natural forests for timber and fuelwood production.

Forest Categories

28 forest categories are used to enable the translation of forest types from national and regional classification systems to a harmonised global one:

Temperate and boreal forest types :

1. Evergreen needleleaf forest – Natural forest with > 30% canopy cover, in which the canopy is predominantly (> 75%) needleleaf and evergreen.
2. Deciduous needleleaf forests – Natural forests with > 30% canopy cover, in which the canopy is predominantly (> 75%) needleleaf and deciduous.
3. Mixed broadleaf/needleleaf forest – Natural forest with > 30% canopy cover, in which the canopy is composed of a more or less even mixture of needleleaf and broadleaf crowns (between 50:50% and 25:75%).
4. Broadleaf evergreen forest – Natural forests with > 30% canopy cover, the canopy being > 75% evergreen and broadleaf.
5. Deciduous broadleaf forest – Natural forests with > 30% canopy cover, in which > 75% of the canopy is deciduous and broadleaves predominate (> 75% of canopy cover).

6. Freshwater swamp forest – Natural forests with > 30% canopy cover, composed of trees with any mixture of leaf type and seasonality, but in which the predominant environmental characteristic is a waterlogged soil.
7. Sclerophyllous dry forest – Natural forest with > 30% canopy cover, in which the canopy is mainly composed of sclerophyllous broadleaves and is > 75% evergreen.
8. Sparse trees and parkland – Natural forests in which the tree canopy cover is between 10–30%, such as in the steppe regions of the world. Trees of any type (e.g., needleleaf, broadleaf, palms).
9. Disturbed natural forest – Any forest type above that has in its interior significant areas of disturbance by people, including clearing, felling for wood extraction, anthropogenic fires, road construction, etc.
10. Exotic species plantation – Intensively managed forests with > 30% canopy cover, which have been planted by people with species not naturally occurring in that country.
11. Native species plantation – Intensively managed forests with > 30% canopy cover, which have been planted by people with species that occur naturally in that country.
12. *Unspecified forest plantation – Forest plantations showing extent only with no further information about their type, This data currently only refers to the Ukraine.
13. *Unclassified forest data – Forest data showing forest extent only with no further information about their type.

Those marked * have been created as a result of data holdings which do not specify the forest type, hence 26 categories are quoted, not 28 shown here.

Tropical Forest Types

1. Lowland evergreen broadleaf rain forest – Natural forests with > 30% canopy cover, below 1,200 m (3,937 ft) altitude that display little or no seasonality, the canopy being >75% evergreen broadleaf.
2. Lower montane forest – Natural forests with > 30% canopy cover, between 1200–1800 m altitude, with any seasonality regime and leaf type mixture.
3. Upper montane forest – Natural forests with > 30% canopy cover, above 1,800 m (5,906 ft) altitude, with any seasonality regime and leaf type mixture.

4. Freshwater swamp forest – Natural forests with > 30% canopy cover, below 1,200 m (3,937 ft) altitude, composed of trees with any mixture of leaf type and seasonality, but in which the predominant environmental characteristic is a waterlogged soil.
5. Semi-evergreen moist broadleaf forest – Natural forests with > 30% canopy cover, below 1,200 m (3,937 ft) altitude in which between 50–75% of the canopy is evergreen, > 75% are broadleaves, and the trees display seasonality of flowering and fruiting.
6. Mixed broadleaf/needleleaf forest – Natural forests with > 30% canopy cover, below 1,200 m (3,937 ft) altitude, in which the canopy is composed of a more or less even mixture of needleleaf and broadleaf crowns (between 50:50% and 25:75%).
7. Needleleaf forest – Natural forest with > 30% canopy cover, below 1,200 m (3,937 ft) altitude, in which the canopy is predominantly (> 75%) needleleaf.
8. Mangroves – Natural forests with > 30% canopy cover, composed of species of mangrove tree, generally along coasts in or near brackish or seawater.
9. Deciduous/semi-deciduous broadleaf forest – Natural forests with > 30% canopy cover, below 1,200 m (3,937 ft) altitude in which between 50–100% of the canopy is deciduous and broadleaves predominate (> 75% of canopy cover).
10. Sclerophyllous dry forest – Natural forests with > 30% canopy cover, below 1,200 m (3,937 ft) altitude, in which the canopy is mainly composed of sclerophyllous broadleaves and is > 75% evergreen.
11. Thorn forest – Natural forests with > 30% canopy cover, below 1,200 m (3,937 ft) altitude, in which the canopy is mainly composed of deciduous trees with thorns and succulent phanerophytes with thorns may be frequent.
12. Sparse trees and parkland – Natural forests in which the tree canopy cover is between 10–30%, such as in the savannah regions of the world. Trees of any type (e.g., needleleaf, broadleaf, palms).
13. Disturbed natural forest – Any forest type above that has in its interior significant areas of disturbance by people, including clearing, felling for wood extraction, anthropogenic fires, road construction, etc.
14. Exotic species plantation – Intensively managed forests with > 30% canopy cover, which have been planted by people with species not naturally occurring in that country.

15. Native species plantation – Intensively managed forests with > 30% canopy cover, which have been planted by people with species that occur naturally in that country.

Forest Loss and Management

The scientific study of forest species and their interaction with the environment is referred to as forest ecology, while the management of forests is often referred to as forestry. Forest management has changed considerably over the last few centuries, with rapid changes from the 1980s onwards culminating in a practice now referred to as sustainable forest management. Forest ecologists concentrate on forest patterns and processes, usually with the aim of elucidating cause and effect relationships. Foresters who practice sustainable forest management focus on the integration of ecological, social and economic values, often in consultation with local communities and other stakeholders. Anthropogenic factors that can affect forests include logging, urban sprawl, human-caused forest fires, acid rain, invasive species, and the slash and burn practices of swidden agriculture or shifting cultivation. The loss and re-growth of forest leads to a distinction between two broad types of forest, primary or old-growth forest and secondary forest. There are also many natural factors that can cause changes in forests over time including forest fires, insects, diseases, weather, competition between species, etc. In 1997, the World Resources Institute recorded that only 20% of the world's original forests remained in large intact tracts of undisturbed forest. More than 75% of these intact forests lie in three countries – the Boreal forests of Russia and Canada and the rainforest of Brazil.

Canada has about 4,020,000 square kilometres (1,550,000 sq mi) of forest land. More than 90% of forest land is publicly owned and about 50% of the total forest area is allocated for harvesting. These allocated areas are managed using the principles of sustainable forest management, which includes extensive consultation with local stakeholders. About eight percent of Canada's forest is legally protected from resource development (Global Forest Watch Canada)(Natural Resources Canada). Much more forest land – about 40 percent of the total forest land base – is subject to varying degrees of protection through processes such as integrated land use planning or defined management areas such as certified forests (Natural Resources Canada).

These maps represent only virgin forest lost. Some regrowth has occurred but not to the age, size or extent of 1620 due to population increases and food cultivation. *From William B. Greeley's, The Relation of Geography to Timber Supply, Economic Geography, 1925, vol. 1, p. 1–11. Source of "Today" map: compiled by George Draffan from roadless area map in The Big Outside: A Descriptive Inventory of the Big Wilderness Areas of the United States, by Dave Foreman and Howie Wolke (Harmony Books, 1992).'*

By December 2006, over 1,237,000 square kilometers of forest land in Canada (about half the global total) had been certified as being sustainably managed (Canadian Sustainable Forestry Certification Coalition). Clearcutting, first used in the latter half of the 20th century, is less expensive, but devastating to the environment and companies are required by law to ensure that harvested areas are adequately regenerated. Most Canadian provinces have regulations limiting the size of clearcuts, although some older clearcuts can range upwards of 110 square kilometres (27,000 acres) in size which were cut over several years. China instituted a ban on logging, beginning in 1998, due to the erosion, and flooding that it caused.

In 2010, the Food and Agriculture Organization of the United Nations reported that world deforestation, mainly the conversion of tropical forests to agricultural land, had decreased over the past ten years but still continues at an alarmingly high rate in many countries. Globally, around 13 million hectares of forests were converted to other uses or lost through natural causes each year between 2000 and 2010 as compared to around 16 million hectares per year during the 1990s. The study covered 233 countries and areas. Brazil and Indonesia, which had the highest loss of forests in the 1990s, have significantly reduced their deforestation rates. In addition, ambitious tree planting programmes in countries such as China, India, the United States and Viet Nam - combined with natural expansion of forests in some regions - have added more than seven million hectares of new forests annually. As a result the net loss of forest area was reduced to 5.2 million hectares per year between 2000 and 2010, down from 8.3 million hectares annually in the 1990s.

In the United States, most forests have historically been affected by humans to some degree, though in recent years improved forestry practices has helped regulate or moderate large scale or severe impacts. However, the United States Forest Service estimates a net loss of about 2 million hectares (4,942,000 acres) between 1997 and 2020; this estimate includes conversion of forest land to other uses, including urban and suburban development, as well as afforestation and natural reversion of abandoned crop and pasture land to forest. However, in many areas of the United States, the area of forest is stable or increasing, particularly in many northern states. The opposite problem from flooding has plagued national forests, with loggers complaining that a lack of thinning and proper forest management has resulted in large forest fires.

Old-growth forest contains mainly natural patterns of biodiversity in established seral patterns, and they contain mainly species native to the region and habitat. The natural formations and processes have not been affected by humans with a frequency or intensity to change the natural structure and components of the habitat. Secondary forest contains significant elements of species which were originally from other regions or habitats. Smaller areas of woodland in cities may be managed as Urban forestry, sometimes within public parks. These are often created for human benefits; Attention Restoration Theory argues that spending time in nature reduces stress and improves health, while forest schools and kindergartens help young people to develop social as well as scientific skills in forests. These typically need to be close to where the children live, for practical logistics.

Greater Yellowstone Ecosystem

The Greater Yellowstone Ecosystem is one of the last remaining large, nearly intact ecosystems in the northern temperate zone of the Earth and is partly located in Yellowstone National Park. Conflict over management has been controversial, and the area is a flagship site among conservation groups that promote ecosystem management. The Greater Yellow Ecosystem (GYE) is one of the world's foremost natural laboratories in landscape ecology and geology and is a world-renowned recreational site. It is also home to the animals of Yellowstone.

History

Yellowstone National Park boundaries were arbitrarily drawn in 1872 in hopes of including all regional geothermal basins in the area. No other landscape considerations were incorporated. By the 1970s, however, the grizzly bear's (*Ursus arctos*) range in and near the park became the first informal minimum boundary of a theoretical Greater Yellowstone Ecosystem that included at least 16,000 square kilometres (4,000,000 acres). Since then, definitions of the greater ecosystem's size have steadily grown larger. A 1994 study listed the size as 76,890 square kilometres (19,000,000 acres), while a 1994 speech by a Greater Yellowstone Coalition leader enlarged that to 80,000 square kilometres (20,000,000 acres).

In 1985 the United States House of Representatives Subcommittees on Public Lands and National Parks and Recreation held a joint subcommittee hearing on Greater Yellowstone, resulting in a 1986 report by the Congressional Research Service outlining shortcomings in interagency coordination and concluding that the area's essential values were at risk.

Other federally managed areas within the GYE include Gallatin, Custer, Caribou-Targhee, Bridger-Teton and Shoshone National Forests, the John D. Rockefeller, Jr. Memorial Parkway, as well as the National Elk Refuge and Grand Teton National Park. The GYE also encompasses some privately held lands surrounding those managed by the U.S. Government. Outside of Yellowstone National Park, ten distinct wilderness areas have been established in the National Forests since 1966 to ensure a higher level of habitat protection than is normally mandated.

Ecosystem Management by Species

The great ecosystem concept has been most often advanced through concerns over individual species rather than over broader ecological principles. Though 20 or 30 or even 50 years of information on a population may be considered long-term by some, one of the important lessons of Greater Yellowstone management is that even half a century is not long enough to give a full idea of how a species may vary in its occupation of a wild ecosystem.

For example, anecdotal information on grizzly bear abundance dates to the mid-19th century, and administrators have made informal population estimates for more than 70 years. From these sources, ecologists know the species was common in Greater Yellowstone when Europeans arrived and that the population was not isolated before the 1930s, but is now. Researchers do not know if bears were more or less common than now. A 1959-1970 bear study suggested a grizzly bear population size of about 176, later revised to about 229. Later estimates have ranged as low as 136 and as high as 540; the most recent is a minimum estimate of 236. Although the Greater Yellowstone population is relatively close to recovery goals, the plan's definition of recovery is controversial. Thus, even though the population may be stable or possibly increasing in the short term, in the longer term, continued habitat loss and increasing human activities may well reverse the trend.

Yellowstone cutthroat trout (*Oncorhynchus clarki bouvieri*) have suffered considerable declines since European settlement, but recently began flourishing in some areas. Especially in Yellowstone Lake itself, long-term records indicate an almost remarkable restoration of robust populations from only three decades ago when the numbers of this fish were depleted because of excessive harvest. Its current recovery, though a significant management achievement, does not begin to restore the species' historical abundance.

Early accounts of pronghorn (*Antilocapra americana*) in Greater Yellowstone described herds of hundreds seen ranging through most major river valleys. These populations were decimated by 1900, and declines continued among remaining herds. On the park's northern range, pronghorn declined from 500-700 in the 1930s to about 122 in 1968. By 1992 the herd had increased to 536. Among plants, whitebark pine (*Pinus albicaulis*) is a species of special interest, in large part because of its seasonal importance to grizzly bears, but also because its distribution could be dramatically reduced by relatively minor global warming. In this case, researchers do not have a good long-term data set on the species, but they understand its ecology well enough to project declining future conservation status. A more immediate, and serious, threat to whitebark pine is an introduced fungal disease, *white pine blister rust* (*Cronartium ribicola*), which is causing heavy mortality in the species. Occasional resistant individuals occur, but in the short to medium term, a severe population decline is expected.

Estimates of the decline of quaking aspen (*Populus tremuloides*) on the park's northern range since 1872 range from 50% to 95%, and perhaps no controversy underway in Greater Yellowstone more clearly reveals the need for comprehensive interdisciplinary research. Several factors are suspected in the aspen's changing status, including Native American influences on numerous mammal species and on fire-return intervals before the creation of the park in 1872; European influences on fire frequency since 1886; regional climate warming; human harvests of beaver and ungulates in the first 15 years of the park's history and of wolves and other predators before 1930; human settlement of traditional ungulate migration routes north of the park since 1872; ungulate (especially elk) effects on all other parts of the ecosystem since 1900; and human influences on elk distribution in the park.

The Yellowstone hot springs are important for their diversity of thermophilic bacteria. These bacteria have been useful in studies of the evolution of photosynthesis and as sources of thermostable enzymes for molecular biology. Although the smell of sulphur is common and there are some sulphur fixing cyanobacteria, it has been found that hydrogen is being used as an energy source by extremophile microbes.

Gray Wolf Reintroduction

The park is a commonly cited example of apex predators affecting an ecosystem through a trophic cascade. After the reintroduction of the gray wolf in 1995, researchers noticed drastic changes occurring. Elk, the primary prey of the gray wolf, became less abundant and changed their behaviour, freeing riparian zones from constant grazing. The respite allowed willows and aspens to grow, creating habitat for beaver, moose, and scores of other species. In addition to the effects on prey species, the gray wolf's presence also affected the park's grizzly bear population. The bears, emerging from hibernation, chose to scavenge off wolf kills to gain needed energy and fatten up after fasting for months. Dozens of other species have been documented scavenging from wolf kills.

Large Marine Ecosystem

Large marine ecosystems (LMEs) are regions of the world's oceans, encompassing coastal areas from river basins and estuaries to the seaward boundaries of continental shelves and the outer margins of the major ocean current systems. They are relatively large regions on the order of 200,000 km^2 or greater, characterized by distinct bathymetry, hydrography, productivity, and trophically dependent populations. The system of LMEs has been developed by the US National Oceanic and Atmospheric Administration (NOAA) to identify areas of the oceans for conservation purposes. The objective is to use the LME concept as a tool for enabling ecosystem-based management to provide a collaborative approach to management of resources within ecologically-bounded transnational areas. This will be done in an international context and consistent with customary international law as reflected in 1982 UN Convention on the Law of the Sea.

LME-based conservation is based on recognition that the world's coastal ocean waters are degraded by unsustainable fishing practices, habitat degradation, eutrophication, toxic pollution, aerosol contamination, and emerging diseases, and that positive actions to mitigate these threats require coordinated actions by governments and civil society to recover depleted fish populations, restore degraded habitats and reduce coastal pollution.

Although the LMEs cover only the continental margins and not the deep oceans and oceanic islands, the 64 LMEs produce 95% of the world's annual marine fishery biomass yields. Most of the global ocean pollution, overexploitation, and coastal habitat alteration occur within their waters. NOAA has conducted studies of principal driving forces affecting changes in biomass yields for 33 of the 64 LMEs, which have been peer-reviewed and published in ten volumes.

Littoral Zone

The littoral zone is the part of a sea, lake or river that is close to the shore. In coastal environments the littoral zone extends from the high water mark, which is rarely inundated, to shoreline areas that are permanently submerged. It always includes this intertidal zone and is often used to mean the same as the intertidal zone. However, the meaning of "littoral zone" can extend well beyond the intertidal zone. There is no single definition. What is regarded as the full extent of the littoral zone, and the way the littoral zone is divided into subregions, varies in different contexts (lakes and rivers have their own definitions). The use of the term also varies from one part of the world to another, and between different disciplines. For example, military commanders speak of the littoral in ways that are quite different from marine biologists.

The adjacency of water gives a number of distinctive characteristics to littoral regions. The erosive power of water results in particular types of landforms, such as sand dunes, and estuaries. The natural movement of the littoral along the coast is called the littoral drift. Biologically, the ready availability of water enables a greater variety of plant and animal life, and particularly the formation of extensive wetlands. In addition, the additional local humidity due to evaporation usually creates a microclimate supporting unique types of organisms.

The word "littoral" is used both as a noun and an adjective. It derives from the Latin noun *litus, litoris*, meaning "shore". (The doubled 't' is a late medieval innovation and the word is sometimes seen in the more classical-looking spelling 'litoral'.)

In Oceanography and Marine Biology

In oceanography and marine biology, the idea of the littoral zone is extended roughly to the edge of the continental shelf. Starting from the shoreline, the littoral zone begins at the spray region just above the high tide mark. From here, it moves to the intertidal region between the high and low water marks, and then out as far as the edge of the continental shelf. These three subregions are called, in order, the supralittoral zone, the eulittoral zone and the sublittoral zone.

Supralittoral Zone

The supralittoral zone (also called the splash, spray, or supratidal zone) is the area above the spring high tide line that is regularly splashed, but not submerged by ocean water. Seawater penetrates these elevated areas only during storms with high tides. Organisms here must cope also with exposure to bad air, fresh water from rain, cold, heat and predation by land animals and seabirds. At the top of this area, patches of dark lichens can appear as crusts on rocks. Some types of periwinkles, Neritidae and detritus feeding Isopoda commonly inhabit the lower supralittoral.

Eulittoral Zone

The eulittoral zone (also called the midlittoral or mediolittoral zone) is the intertidal zone, also known as the foreshore. It extends from the spring high tide line, which is rarely inundated, to the neap low tide line, which is rarely not inundated.

The wave action and turbulence of recurring tides shapes and reforms cliffs, gaps, and caves, offering a huge range of habitats for sedentary organisms. Protected rocky shorelines usually show a narrow almost homogenous eulittoral strip, often marked by the presence of barnacles. Exposed sites show a wider extension and are often divided into further zones.

Sublittoral Zone

The sublittoral zone, also called the neritic zone, starts immediately below the eulittoral zone. This zone is permanently covered with seawater. In physical oceanography, the sublittoral zone refers to coastal regions with significant tidal flows and energy dissipation, including non-linear flows, internal waves, river outflows and oceanic fronts. In practice, this typically extends to the edge of the continental shelf, with depths around 200 meters. In marine biology, the sublittoral refers to the areas where sunlight reaches the ocean floor, that is, where the water is never so deep as to take it out of the photic zone.

This results in high primary production and makes the sublittoral zone the location of the majority of sea life. As in physical oceanography, this zone typically extends to the edge of the continental shelf. The benthic zone in the sublittoral is much more stable than in the intertidal zone; temperature, water pressure, and the amount of sunlight remain fairly constant. Sublittoral corals do not have to deal with as much change as intertidal corals. Corals can live in both zones, but they are more common in the sublittoral zone.

Within the sublittoral, marine biologists also identify the following:

- The infralittoral zone is the algal dominated zone to maybe five metres below the low water mark.

- The circalittoral zone is the region beyond the infralittoral, that is, below the algal zone and dominated by sessile animals such as oysters.

Shallower region of the sublittoral zone, extending not far from the shore, are sometimes referred to as the subtidal zone.

In Freshwater Ecosystems

In freshwater situations, littoral zones occur on the edge of large lakes and rivers, often with extensive areas of wetland. Hence, they are sometimes referred to as fringing wetlands. Here, the effects of tides are minimal, so other definitions of "littoral" are used. For example, the Minnesota Department of Natural Resources defines littoral as that portion of the lake that is less than 15 feet in depth.

The littoral zone may form a narrow or broad fringing wetland, with extensive areas of aquatic plants sorted by their tolerance to different water depths. Typically, four zones are recognized, from higher to lower on the shore: wooded wetland, wet meadow, marsh and aquatic vegetation. The relative areas of these four types depends not only on the profile of the shoreline, but upon past water levels. The area of wet meadow is particularly dependent upon past water levels; in general, the area of wet meadows along lakes and rivers increases with natural water level fluctuations. Many of the animals in lakes and rivers are dependent upon the wetlands of littoral zones, since the rooted plants provide habitat and food. Hence, a large and productive littoral zone is considered an important characteristic of a healthy lake or river.

Littoral zones are at particular risk for two reasons. First, human settlement is often attracted to shorelines, and settlement often disrupts breeding habitats for littoral zone species. For example, many turtles are killed on roads when they leave the water to lay their eggs in upland sites. Fish can be negatively affected by docks and retaining walls which remove breeding habitat in shallow water. Some shoreline communities even deliberately try to remove wetlands since they may interfere with activities like swimming. Overall, the presence of human settlement has a demonstrated negative impact upon adjoining wetlands. An equally serious problem is the tendency to stabilize lake or river levels with dams. Dams removed the spring flood which carries nutrients into littoral zones, and reduces the natural fluctuation of water levels upon which many wetland plants and animals depend. Hence, over time, dams can reduce the area of wetland from a broad littoral zone to a narrow band of vegetation. Marshes and wet meadows are at particular risk.

Other Definitions

For the purposes of naval operations, the United States Navy divides the littoral zone in the ways shown on the diagram at the top of this article. The United States Army Corps of Engineers and Environmental Protection Agency have their own definitions, and these have legal implications.

Riparian Zone

A riparian zone or riparian area is the interface between land and a river or stream. Riparian is also the proper nomenclature for one of the fifteen terrestrial biomes of the earth. Plant habitats and communities along the river margins and banks are called riparian vegetation, characterized by hydrophilic plants. Riparian zones are significant in ecology, environmental management, and civil engineering because of their role in soil conservation, their habitat biodiversity, and the influence they have on fauna and aquatic ecosystems, including grassland, woodland, wetland or even non-vegetative. In some regions the terms riparian woodland, riparian forest, riparian buffer zone, or riparian strip are used to characterize a riparian zone. The word "riparian" is derived from Latin *ripa*, meaning river bank.

Characteristics

Riparian zones may be natural or engineered for soil stabilization or restoration. These zones are important natural biofilters, protecting aquatic environments from excessive sedimentation, polluted surface runoff and erosion. They supply shelter and food for many aquatic animals and shade that is an important part of stream temperature regulation. When riparian zones are damaged by construction, agriculture or silviculture, biological restoration can take place, usually by human intervention in erosion control and revegetation. If the area adjacent to a watercourse has standing water or saturated soil for as long as a season, it is normally termed a wetland because of its hydric soil characteristics. Because of their prominent role in supporting a diversity of species, riparian zones are often the subject of national protection in a Biodiversity Action Plan. These also known as a "Plant or Vegetation Waste Buffer".

Research shows riparian zones are instrumental in water quality improvement for both surface runoff and water flowing into streams through subsurface or groundwater flow. Particularly the attenuation of nitrate or denitrification of the nitrates from fertilizer in this buffer zone is important. Riparian zones can play a role in lowering nitrate contamination in surface runoff from agricultural fields, which runoff would otherwise damage ecosystems and human health. The use of wetland riparian zones shows a particularly high rate of removal of nitrate entering a stream and thus has a place in agricultural management.

Roles and Functions

Riparian zones dissipate stream energy. The meandering curves of a river, combined with vegetation and root systems, dissipate stream energy, which results in less soil erosion and a reduction in flood damage. Sediment is trapped, reducing suspended solids to create less turbid water, replenish soils, and build stream banks. Pollutants are filtered from surface runoff which enhances water quality via biofiltration.

The riparian zones also provide wildlife habitat, increased biodiversity, and provide wildlife corridors, enabling aquatic and riparian organisms to move along river systems avoiding isolated communities. They can provide forage for wildlife and livestock.

They provide native landscape irrigation by extending seasonal or perennial flows of water. Nutrients from terrestrial vegetation (e.g. plant litter and insect drop) is transferred to aquatic food webs. The vegetation surrounding the stream helps to shade the water, mitigating water temperature changes. The vegetation also contributes wood debris to streams which is important to maintaining geomorphology.

From a social aspect, riparian zones contribute to nearby property values through amenity and views, and they improve enjoyment for footpaths and bikeways through supporting foreshoreway networks. Space is created for riparian sports including fishing, swimming and launching for vessels and paddlecraft.

The riparian zone acts as a sacrificial erosion buffer to absorb impacts of factors including climate change, increased runoff from urbanisation and increased boatwake without damaging structures located behind a setback zone.

Role in Logging

The protection of riparian zones is often a consideration in logging operations. The undisturbed soil, soil cover, and vegetation provide shade, plant litter, woody material, and reduce the delivery of soil eroded from the harvested area. Factors such as soil types and root structures, climatic conditions and above ground vegetative cover impact the effectiveness of riparian buffering.

Vegetation

The assortment of riparian zone trees varies from those of wetlands and typically consists of plants that either are emergent aquatic plants, or herbs, trees and shrubs that thrive in proximity to water.

North America

Water's edge: Herbaceous Perennial:

- *Peltandra virginica* - Arrow Arum
- *Sagittaria lancifolia* - Arrowhead
- *Carex stricta* - Tussock Sedge
- *Iris virginica* - Southern Blue Flag Iris

Inundated Riparian Zone

Herbaceous Perennial:

- *Sagittaria latifolia* - Duck Potato
- *Schoenoplectus tabernaemontani* - Softstem Bulrush
- *Scorpus americanus* - Three-square Bulrush
- *Eleocharis quadrangulata* - Square-stem Spikerush
- *Eleocharis obusa* - Spikerush

Western

In western North America and the Pacific Coast the riparian vegetation includes: Riparian trees

- *Sequoia sempervirens* - Coast Redwood
- *Thuja plicata* - Western Redcedar
- *Abies grandis* - Grand Fir
- *Picea sitchensis* - Sitka Spruce
- *Chamaecyparis lawsoniana* - Port Orford-cedar
- *Taxus brevifolia* - Pacific Yew
- *Populus fremontii* - Fremont Cottonwood
- *Populus trichocarpa* - Black Cottonwood
- *Platanus racemosa* - California Sycamore
- *Alnus rhombifolia* - White Alder
- *Alnus rubra* - Red Alder
- *Acer macrophyllum* - Big-leaf Maple
- *Fraxinus latifolia* - Oregon ash
- *Prunus emarginata* - Bitter Cherry

- *Salix lasiolepis* - Arroyo Willow
- *Salix lucida* - Pacific Willow
- *Quercus agrifolia* - Coast live oak
- *Quercus garryana* - Garry oak
- *Populus tremuloides* - Quaking Aspen
- *Umbellularia californica* - California Bay Laurel
- *Cornus nuttallii* - Pacific Dogwood

Riparian shrubs:

- *Acer circinatum* - Vine Maple
- *Ribes spp.* - Gooseberies and Currants
- *Rosa pisocarpa* - Swamp Rose or Cluster Rose
- *Symphoricarpos albus* - Snowberry
- *Spirea douglasii* - Douglas spirea
- *Rubus spp.* - Blackberries, Raspberries, Thimbleberry, Salmonberry
- *Rhododendron occidentale* - Western Azalea
- *Oplopanax horridus* - Devil's Club
- *Oemleria cerasiformis* - Indian Plum, Osoberry
- *Lonicera involucrata* - Twinberry
- *Cornus stolonifera* - Red-osier Dogwood
- *Salix spp.* - Willows

Other plants:

- *Polypodium* - Polypody Ferns
- *Polystichum* - Sword Ferns
- *Woodwardia* - Giant Chain Ferns
- *Pteridium* - Goldback Ferns
- *Dryopteris* - Wood Ferns
- *Adiantum* - Maidenhair Ferns
- *Carex spp.* - Sedges
- *Juncus spp.* - Rushes
- *Festuca californica* - California Fescue bunchgrass
- *Leymus condensatus* - Giant Wildrye bunchgrass
- *Melica californica* - California Melic bunchgrass
- *Mimulus spp.* - Monkeyflower and varieties
- *Aquilegia spp.* - Columbine

Asia

In Asia there are different types of riparian vegetation, but the interactions between hydrology and ecology are similar as occurs in other geographic areas.

- *Carex spp.* - Sedges
- *Juncus spp.* - Rushes

Australia

Typical riparian vegetation in Temperate New South Wales, Australia include:

- *Acacia melanoxylon* - Blackwood
- *Acacia pravissima* - Ovens Wattle
- *Acacia rubida* - Red Stem Wattle
- *Bursaria lasiophylla* - Blackthorn
- *Callistemon citrinus* - Crimson Bottlebrush
- *Callistemon sieberi* - River Bottlebrush
- *Casuarina cunninghamiana* - River She-Oak
- *Eucalyptus bridgesiana* - Apple Box
- *Eucalyptus camaldulensis* - River Red Gum
- *Eucalyptus melliodora* - Yellow Box
- *Eucalyptus viminalis* - Manna Gum
- *Kunzea erocoides* - Burgan
- *Leptospernum obovatum* - River Tea-Tree
- *Melaleuca ericifolia* - Swamp Paperbark

Central Europe

Typical riparian zone trees in Central Europe include:

- *Acer campestre* - Field Maple
- *Acer pseudoplatanus* - Sycamore Maple
- *Alnus glutinosa* - Black Alder
- *Carpinus betulus* - European Hornbeam
- *Fraxinus excelsior* - European Ash
- *Juglans regia* - Persian Walnut
- *Malus sylvestris* - European Wild Apple
- *Populus alba* - White Poplar
- *Populus nigra* - Black Poplar

- *Quercus robur* - Pedunculate Oak
- *Salix alba* - White Willow
- *Salix fragilis* - Crack Willow
- *Tilia cordata* - Small-leaved Lime
- *Ulmus laevis* - European White Elm
- *Ulmus minor* - Field Elm

Repair and Restoration

Land clearing followed by floods can quickly erode a riverbank, taking valuable grasses and soils downstream, and allowing the sun to bake the land dry. Natural Sequence Farming techniques have been used in the Upper Hunter Valley of New South Wales, Australia to rapidly restore eroded farms to optimum productivity.. The Natural Sequence Farming technique involves placing obstacles in the water's pathway to lessen the energy of a flood, and help the water to deposit soil and seep into the flood zone. Another technique is to quickly establish ecological succession by encouraging fast growing plants such as "weeds" (pioneer species) to grow. These can quickly stabilize the soil, place carbon into the ground, and protect the land from drying. The weeds will improve the streambeds so that trees and grasses can return, and later replace the weeds.

Subsurface Lithoautotrophic Microbial Ecosystem

Subsurface lithoautotrophic microbial ecosystems, or "SLIMEs" (also abbreviated "SLMEs" or "SLiMEs"), are defined by Edward O. Wilson as "unique assemblages of bacteria and fungi that occupy pores in the interlocking mineral grains of igneous rock beneath Earth's surface."

Urban Ecosystem

Urban ecosystems are the cities, towns, and urban strips constructed by humans. This is the growth in the urban population and the supporting built infrastructure has affected both urban environments and also on areas which surround urban areas. These include semi or 'peri-urban' environments that fringe cities as well as agricultural and natural landscapes. Scientists are now developing ways to measure and understand the effects of urbanisation on human and environmental health.

By considering urban areas as part of a broader ecological system, scientists can investigate how urban landscapes function and how they affect other landscapes with which they interact. In this context, urban environments are affected by their surrounding environment but also affect that environment. Knowing this may provide clues as to which alternative development options will lead to the best overall environmental outcome.

CSE's urban ecosystem research is focused on:

- Understanding how cities work as ecological system
- Developing sustainable approaches to development of city fringe areas that reduce negative impact on surrounding environments
- Developing approaches to urban design that provide for health and opportunity for citizens.

Movile Cave

Movile Cave is a cave in Constanþa County, Romania discovered by Cristian Lascu in 1986 a few kilometers from the Black Sea coast. It is notable for its unique groundwater ecosystem rich in hydrogen sulfide and carbon dioxide but very poor in oxygen.

Life in the cave has been separated from the outside for the past 5.5 million years and it is based completely on chemosynthesis rather than photosynthesis.

The air in the cave is very different from the outer atmosphere. The level of oxygen is only a third to half of the concentration found in open air (7-10% O_2 in the cave atmosphere, compared to 21% O_2 in air), and about one hundred times more carbon dioxide (2-3.5% CO_2 in the cave atmosphere, versus 0.03% CO_2 in air). It also contains 1-2% methane (CH_4) and both the air and waters of the Cave contain high concentrations of hydrogen sulfide (H_2S) and ammonia (NH_3).

Forty-eight species, among them leeches, spiders, scorpions and insects, were found inside the cave, of which 33 are endemic. The food chain is based on chemosynthesis in the form of methane and sulphur oxidising bacteria, which in turn release nutrients for fungi and other bacteria. This forms microbial mats on the cave walls and the surface of lakes and ponds that is grazed on by some of the animals. The grazers are then preyed on by predatory species.

Desert

A desert is a landscape or region of land that is very dry because of low rainfall amounts (precipitation), often has little coverage by plants, and in which streams dry up unless they are supplied by water from outside areas. Deserts can also be described as areas where more water is lost by evapotranspiration than falls as precipitation. Desert plants must have special adaptations to survive with this little water. Deserts generally receive less than 250 millimetres (10 in) of rain (precipitation) each year. Semideserts or steppes are regions which receive between 250 millimetres (10 in) and 400 to 500 millimetres (16 to 20 in).

Physical Geography

Classification: Deserts have been defined and classified in a number of ways, generally combining total rainfall, number of days on which rain falls, temperature, and humidity, and sometimes additional factors. For example, Phoenix, Arizona receives less than 250 millimeters (10 in) of precipitation per year, and is immediately recognized as being located in a desert due to its arid adapted plants. The North Slope of Alaska's Brooks Range also receives less than 250 millimeters (10 in) of precipitation per year and is often classified as a cold desert. Other regions of the world have cold deserts, including areas of the Himalayas and other high altitude areas in other parts of the world. Polar deserts cover much of the ice free areas of the Arctic and Antarctic. A non-technical definition is that deserts are those parts of the Earth's surface that have insufficient vegetation cover to support a human population.

Potential evapotranspiration supplements the measurement of rainfall in providing a scientific measurement-based definition of a desert. The water budget of an area can be calculated using the formula $P - PE \pm S$, wherein P is precipitation, PE is potential evapotranspiration rates and S is amount of surface storage of water. Evapotranspiration is the combination of water loss through atmospheric evaporation and through the life processes of plants. Potential evapotranspiration, then, is the amount of water that *could* evaporate in any given region. As an example, Tucson, Arizona receives about 300 millimeters (12 in) of rain per year, however about 2,500 millimeters (100 in) of water could evaporate over the course of a year. In other words, about eight times more water could evaporate from the region than actually falls as rain. Rates of evapotranspiration in cold regions such as Alaska are much lower because of the lack of heat to aid in the evaporation process. Deserts are sometimes classified as "hot" or "cold", "semiarid" or "coastal". Hot deserts are mostly located near the Tropics of Capricorn or Cancer. The mean temperature is about 22 °C (72 °F), with a daily range of 40 °C (104 °F) to 7 °C (45 °F) or even lower. The rainfall is very low, especially in winter, and may come in the form of a very occasional heavy downpour. The soil is coarse-textured gravel or sand, shallow and well drained. Plants here tend to have deep taproots and may only open their stomata at night.

Cold deserts can be covered with snow or ice for part of the year; frozen water unavailable to plant life. They are found in Greenland, the nearctic ecozone of North America and Antarctica. The mean winter temperature is between 4 °C (39 °F) and "2 °C (28 °F) and the annual precipitation between 15 and 26 cm (6 and 10 in). The soil is fine silt, saline and heavy. Plants growing here tend to be widely separated, deciduous, low and spiny.

Semiarid deserts have long, mostly dry summers and little rainfall in winter. The temperature does not rise as high as in hot deserts, averaging 21 °C (70 °F) to 27 °C (81 °F) in summer and the evenings and nights are cool. As a result, condensation occurs and these areas may receive most of their precipitation in the form of dew. The soil ranges from a fine sand to a coarse gravel and is usually shallow and well-drained. Plants tend to be spiny or have glaucous, silvery or glossy leaves.

Coastal deserts are mostly found on the western edges of continental land masses in regions where cold currents from the polar regions approach the land or cold water upwellings rise from the ocean depths. The cool winds crossing this water pick up little moisture and the coastal regions receive only slight precipitation though they often experience mists or fogs, especially in the winter. The average summer temperature is in the range 13 °C (55 °F) to 24 °C (75 °F) and the winter temperature lower than 6 °C (43 °F). The soil is usually fine-textured and well drained. Deserts of this type occur in Chile, south-west Africa, southern California and Baja California.

In 1961, Peveril Meigs divided desert regions on Earth into three categories according to the amount of precipitation they received. In this now widely accepted system, extremely arid lands have at least twelve consecutive months without rainfall, arid lands have less than 250 mm (10 in) of annual rainfall, and semiarid lands have a mean annual precipitation of between 250 and 500 mm (10–20 in). Both extremely arid and arid lands are considered to be deserts while semiarid lands are generally referred to as steppes.

Deserts are also classified by their geographical location and dominant weather pattern as trade wind, mid-latitude, rain shadow, coastal, monsoon, or polar deserts. Trade wind deserts occur either side of the horse latitudes at 30° to 35° North and South. These belts are associated with the subtropical anticyclone and the large-scale descent of dry air moving from high-altitudes toward the poles. The Sahara Desert is of this type. Mid-latitude deserts occur between 30°

and 50° North and South. They are mostly in areas remote from the sea where most of the moisture has already precipitated from the prevailing winds. They include the Tengger and Sonoran deserts. Monsoon deserts are similar. They occur in regions where large temperature differences occur between sea and land. Moist warm air rises over the land, deposits its water content and circulates back to sea. Further inland, areas receive very little precipitation. The Thar Desert near the India/Pakistan border is of this type.

In some parts of the world, deserts are created by a rain shadow effect. Orographic lift occurs as air masses rise to pass over high ground. In the process they cool and lose much of their moisture by precipitation on the windward slope of the mountain range. When they descend on the leeward side, they warm and their capacity to hold moisture increases so an area with relatively little rainfall occurs. The Taklamakan Desert is an example, lying in the rain shadow of the Himalayas and receiving less than 38 millimetres (1.5 in) rainfall annually. Other areas are arid by virtue of being a very long way from the nearest available sources of moisture.

Montane deserts are arid places with a very high altitude; the most prominent example is found north of the Himalayas, in the Kunlun Mountains and the Tibetan Plateau. Many locations within this category have elevations exceeding 3,000 meters (10,000 ft) and the thermal regime can be hemiboreal. These places owe their profound aridity (the average annual precipitation is often less than 40 mm or 1.5 in) to being very far from the nearest available sources of moisture and are often in the lee of mountain ranges. Montane deserts are normally cold, or may be scorchingly hot by day and very cold by night as is true of the northeastern slopes of Mount Kilimanjaro. Polar deserts such as McMurdo Dry Valleys remain ice-free because of the dry katabatic winds that flow downhill from the surrounding mountains. Former desert areas presently in non-arid environments, such as Sandhills (Nebraska), are known as paleodeserts. In the Köppen climate classification system, deserts are classed as *BWh* (hot desert) or *BWk* (temperate desert). In the Thornthwaite climate classification system, deserts would be classified as arid megathermal climates.

Weathering Processes

Deserts usually have a large diurnal and seasonal temperature range, with high daytime temperatures, and low nighttime temperatures. The humidity may be as low as 2 to 5% and because water vapour in the atmosphere acts to trap long wave infrared radiation from the ground, the cloudless desert sky is incapable of blocking sunlight during the day or trapping heat during the night. Thus, during daylight, most of the sun's heat reaches the ground, and as soon as the sun sets the desert cools quickly by radiating its heat into space. In hot deserts, the temperature in the daytime can exceed 45 °C (113 °F) in the summer and dip to below freezing point at night during the winter.

Such large temperature variations have a destructive effect on the exposed rocky surfaces. The repeated fluctuations put a strain on exposed rock and the flanks of mountains crack and shatter. Fragmented strata slide down into the valleys where they continue to break into pieces due to the remorseless sun by day and chill by night. Successive strata are exposed to further weathering. The relief of the internal pressure that has built up in rocks that have been underground for aeons can cause them to shatter. Exfoliation also occurs when the outer surfaces of rocks split off in flat flakes. This is caused by moisture that has entered through minute cracks becoming trapped. Minerals present in the rock in the form of salts and clay swell up and a strain develops which causes a thin layer of stone to break off.

As the desert mountains decay, large areas of shattered rock and rubble occur. The process continues and the end products are either dust or sand. Dust is formed from solidified clay or volcanic deposits whereas sand results from the fragmentation of harder granites, limestone and sandstone. There is a certain critical size (about 500μ) below which further temperature-induced weathering of rocks does not occur and this provides a minimum size for sand grains. As the mountains are eroded, more and more sand is created. At high wind speeds, sand grains are picked up off the surface and blown along, a process known as saltation. The whirling airborne grains act as a sand blasting mechanism which grinds away solid objects in its path as the kinetic energy of the wind is transferred to the ground. The sand eventually ends up deposited in level areas known as sand-fields or sand-seas, or piled up in dunes.

Dust Storms and Sandstorms

Sand and dust storms are natural events that occur in arid regions where the land is not protected by a covering of vegetation. Dust storms usually start in desert margins rather than the deserts themselves where the finer materials have already been blown away. As a steady wind begins to blow, fine particles lying on the exposed ground begin to vibrate.

At greater wind speeds, some particles are lifted into the air stream. When they land, they strike other particles which may be jerked into the air in their turn, starting a chain reaction. Once ejected, these particles move in one of three possible ways, depending on their size and density; suspension, saltation or creep. Suspension is only possible for particles less than 0.1 mm (0.004 in) in diameter.

In a dust storm, these fine particles are lifted up and wafted aloft to heights of up to 6 kilometres (3.7 mi). They reduce visibility and can remain in the atmosphere for days on end, conveyed by the trade winds distances of thousands of kilometres (miles). Denser clouds of dust can be formed in stronger winds, moving across the land with a billowing leading edge.

The sunlight can be obliterated and it may become as dark as night at ground level. In a study of a dust storm in China in 2001, it was estimated that 6.5 million tons of dust were involved, covering an area of 134,000,000 square kilometres (52,000,000 sq mi). The mean particle size was 1.44 μm.

Sandstorms occur with much less frequency. They are often preceded by severe dust storms and occur when the wind velocity increases to a point where it can lift heavier particles. These grains of sand, up to about 0.5 millimetres (0.020 in) in diameter are jerked into the air but soon fall back to earth, ejecting other particles in the process. Their weight prevents them from being airborne for long and most only travel a distance of a few metres (yards). The sand streams along above the surface of the ground like a fluid, often rising to heights of about 30 centimetres (12 in). In a really severe steady blow, 2 metres (6 ft 7 in) is about as high as the sand stream can rise as the largest sand grains do not become airborne at all. They are transported by creep, being rolled along the desert floor or performing short jumps.

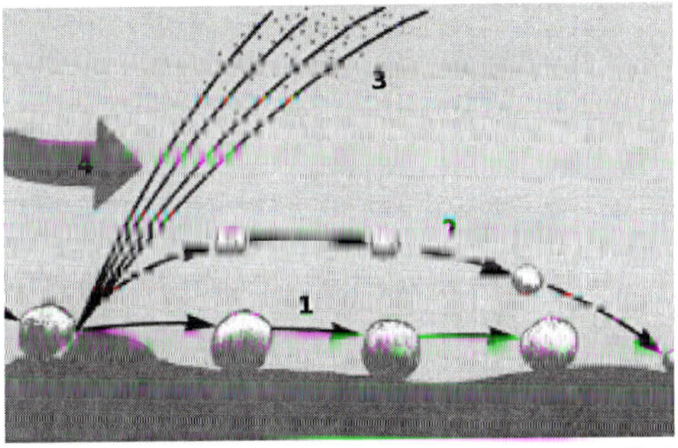

Figure: *Wind-blown particles:* 1=*Creep* 2=*Saltation* 3=*Suspension* 4=*Wind current*

During a sandstorm, the wind-blown sand particles become electrically charged. The electric field can produce sparks and interference with telecommunications equipment and range in size between 1 and 200 kV/m. The electric fields are caused by collision between airborne particles and when saltating sand grains impact the ground. The mechanism is little understood but the particles usually have a negative charge when their diameter is under 250 μm and a positive one when they are over 500 μm.

Major Deserts

Deserts take up about one third (33%) of the Earth's land surface. Bottomlands may be salt-covered flats. Eolian processes are major factors in shaping desert landscapes. Polar deserts (also seen as "cold deserts") have similar features, except the main form of precipitation is snow rather than rain. Antarctica is the world's largest cold desert (composed of about 98% thick continental ice sheet and 2% barren rock). Some of the barren rock is to be found in the so-called Dry Valleys of Antarctica that almost never get snow, which can have ice-encrusted saline lakes that suggest evaporation far greater than the rare snowfall due to the strong katabatic winds that evaporate even ice.

The Ten Largest Deserts

Rank	*Desert*	*Area (km²)*	*Area (mi²)*
1.	Antarctic Desert (Antarctica)	14,200,000	5,500,000
2.	Arctic Desert (Arctic)	13,900,000	5,400,000
3.	Sahara Desert (Africa)	9,100,000	3,500,000
4.	Arabian Desert (Middle East)	2,600,000	1,000,000
5.	Gobi Desert (Asia)	1,300,000	500,000
6.	Patagonian Desert (South America)	670,000	260,000
7.	Great Victoria Desert (Australia)	647,000	250,000
8.	Kalahari Desert (Africa)	570,000	220,000
9.	Great Basin Desert (North America)	490,000	190,000
10.	Thar Desert (India, Pakistan)	450,000	175,000

Deserts, both hot and cold, play a part in moderating the Earth's temperature. This is because they reflect more of the incoming light and their albedo is higher than that of forests or the sea.

Features

Contrary to the impression given by films such as *Desert Sands* (1955) or *Ice Cold in Alex* (1958), and books such as Wilfred Thesiger's

Arabian Sands, deserts do not always consist of sand. Across the world, around 20 percent of desert is sand, varying from only two percent in North America to 30% in Australia and over 45% in Central Asia. Where sand occurs, it is usually in large quantities consisting of sand sheets or extensive areas of dunes.

A sand sheet is a near-level, firm expanse of partially-consolidated particles in a layer that varies from a few centimetres to a few metres thick. The structure of the sheet consists of thin horizontal layers of coarse silt and very fine to medium grain sand, separated by layers of coarse sand and pea-gravel which are a single grain thick. These larger particles anchor the other particles in place and may also be packed together on the surface so as to form a miniature desert pavement. Small ripples form on the sand sheet when the wind exceeds 24 kph (15 mph). They form perpendicular to the wind direction and gradually move across the surface as the wind continues to blow. The distance between their crests corresponds to the average length of jumps made by particles during saltation. The ripples are ephemeral and a change in wind direction causes them to reorganise.

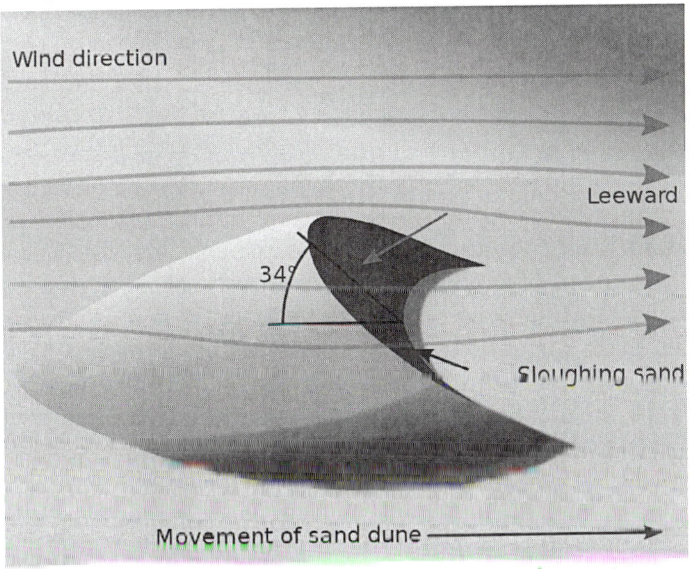

***Figure:** Diagram showing barchan dune formation, with the wind blowing from the left*

Sand dunes are accumulations of windblown sand piled up in mounds or ridges. They form downwind of copious sources of dry, loose sand and occur when topographic and climatic conditions cause airborne particles to settle. As the wind blows, saltation and creep take place on the windward side of the dune and individual grains of sand move uphill. When they reach the crest, they cascade down the far side. The upwind slope typically has a gradient of 10° to 20° while the lee slope is around 32°, the angle at which loose dry sand will slip. As this wind-induced movement of sand grains takes place, the dune moves slowly across the surface of the ground. Dunes are sometimes solitary, but they are more often grouped together in dune fields. When these are extensive, they are known as sand seas or ergs.

The shape of the dune depends on the characteristics of the prevailing wind. Barchan dunes are produced by strong winds blowing across a level surface, and are crescent-shaped with the concave side away from the wind. When there are two directions from which winds regularly blow, a series of long, linear dunes known as seif dunes may form. These also occur parallel to a strong wind that blows in one general direction. Transverse dunes run at a right angle to the prevailing wind direction. Star dunes are formed by variable winds, and have several ridges and slip faces radiating from a central point. They tend to grow vertically; they can reach a height of 500 metres (1,600 ft), making them the tallest type of dune. Rounded mounds of sand without a slip face are the rare dome dunes, found on the upwind edges of sand seas.

A large part of the surface area of the world's deserts consists of flat, stone-covered plains dominated by wind erosion. In "eolian deflation", the wind continually removes fine-grained material, which becomes wind-blown sand. This exposes coarser-grained material, mainly pebbles with some larger stones or cobbles, leaving a desert pavement, an area of land overlaid by closely packed smooth stones forming a tessellated mosaic. Different theories exist as to how exactly the pavement is formed. It may be that after the sand and dust is blown away by the wind the stones jiggle themselves into place; alternatively, stones previously below ground may in some way work themselves to the surface. Very little further erosion takes place after the formation of a pavement, and the ground becomes stable. Evaporation brings moisture to the surface by capillary action and calcium salts may be precipitated, binding particles together to form a desert conglomerate. In time, bacteria that live on the surface of the stones accumulate a film of minerals and clay particles, forming a shiny brown coating known as desert varnish.

Other non-sandy deserts consist of exposed outcrops of bedrock, dry soils or aridisols, and a variety of landforms affected by flowing water, such as alluvial fans, sinks or playas, temporary or permanent lakes, and oases. A hamada is a type of desert landscape consisting of a high rocky plateau where the sand has been removed by aeolian processes. Other landforms include plains largely covered by gravels and angular boulders, from which the finer particles have been stripped by the wind. These are called "reg" in the western Sahara, "serir" in the eastern Sahara, "gibber plains" in Australia and "saï" in central Asia. The Tassili Plateau in Algeria is an impressive jumble of eroded sandstone outcrops, canyons, blocks, pinnacles, fissures, slabs and ravines. In some places the wind has carved holes or arches and in others it has created mushroom-like pillars narrower at the base than the top. In the Colorado Plateau it is water that has been the eroding force. Here the Colorado River has cut its way over the millennia through the high desert floor creating a canyon that is over a mile (6,000 feet or 1,800 metres) deep in places, exposing strata that are over two billion year old.

Water

The driest place on Earth is the Atacama Desert. It is virtually devoid of life because it is blocked from receiving precipitation by the Andes mountains to the east and the Chilean Coast Range to the west. The cold Humboldt Current and the anticyclone of the Pacific are essential to keep the dry climate of the Atacama. The average rainfall in the Chilean region of Antofagasta is just 1 mm per year. Some weather stations in the Atacama have never received rain. Evidence suggests that the Atacama may not have had any significant rainfall from 1570 to 1971. It is so arid that mountains that reach as high as 6,885 meters (22,590 feet) are completely free of glaciers and, in the southern part from 25°S to 27°S, may have been glacier-free throughout the Quaternary, though permafrost extends down to an altitude of 4,400 meters and is continuous above 5,600 meters. Nevertheless, there is some plant life in the Atacama, in the form of specialist plants that obtain moisture from dew and the fogs that blow in from the Pacific.

When rain falls in deserts, as it occasionally does, it is often with great violence. The desert surface is evidence of this with dry stream channels known as arroyos or wadis meandering across its surface. These can experience flash floods, becoming raging torrents with surprising rapidity after a storm that may be many kilometres away. Most deserts are in basins with no drainage to the sea but some are crossed by exotic rivers sourced in mountain ranges or other high rainfall areas beyond their borders. The River Nile, the Colorado

River and the Yellow River do this, losing much of their water through evaporation as they pass through the desert and raising groundwater levels nearby. There may also be underground sources of water in deserts in the form of springs, aquifers, underground rivers or lakes. Where these lie close to the surface, oases may form and plant and animal life flourish. Kharga Oasis in Egypt is 150 kilometres (93 mi) long and results from the removal by wind erosion of sand and dust to expose the water table just above the bedrock.

Lakes may form in basins where there is sufficient rainfall or meltwater from glaciers above. They are usually shallow and saline, and wind blowing over their surface can cause stress, moving the water over nearby low-lying areas. When the lakes dry up, they leave a crust or hardpan behind. This area of deposited clay, silt or sand is known as a playa. The deserts of North America have more than one hundred playas, many of them relics of Lake Bonneville which covered parts of Utah, Nevada and Idaho during the last ice age when the climate was colder and wetter. These include the Great Salt Lake, Utah Lake, Sevier Lake and many dry lake beds. The smooth flat surfaces of playas have been used for attempted vehicle speed records at Black Rock Desert and Bonneville Speedway and the United States Air Force uses Rogers Dry Lake in the Mojave Desert as runways for aircraft and the space shuttle.

Biogeography

Flora: Plants face severe challenges in arid environments. Problems they need to solve include how to obtain enough water, how to avoid being eaten and how to reproduce. Photosynthesis is the key to plant growth. It can only takes place during the day as energy from the sun is required, but during the day, many deserts become very hot. Opening stomata to allow in the carbon dioxide necessary for the process causes evapotranspiration, and conservation of water is a top priority for desert vegetation. Some plants have resolved this problem by adopting crassulacean acid metabolism or C4 carbon fixation. Both these mechanisms allow carbon dioxide to enter the plant at night and be stored inside the tissues for use during the daytime, avoiding the necessity for the stomata to be opened in the heat of the day.

Many desert plants have reduced the size of their leaves or abandoned them altogether. Cacti are desert specialists, the leaves have been dispensed with and the chlorophyll displaced into the trunks, the cellular structure of which have been modified to allow them to store water. When rain falls, the water is rapidly absorbed by the shallow roots and retained to allow them to survive until the next downpour which may be months or years away. The giant saguaro cacti of the Sonoran Desert form "forests", provide shade for other plants and nesting places for desert birds. Saguaro grow slowly but may live for up to two hundred years. The surface of the trunk is folded like a concertina allowing it to expand and a large specimen can hold eight tons of water after a good downpour.

Cacti are restricted to the New World but other xerophytic plants have developed similar strategies by a process known as convergent evolution. They limit water loss by reducing the size and number of stomata, by having waxy coatings and hairy or tiny leaves. Some are deciduous, shedding their leaves in the driest season, and others curl their leaves up to reduce transpiration. Others store water in succulent leaves or stems or in fleshy tubers. Desert plants maximise water uptake by having shallow roots that spread widely, or by developing long taproots that reach down to deep rock strata for ground water. The saltbush in Australia has succulent leaves and secretes salt crystals, enabling it to live in saline areas. In common with cacti, many have developed spines to ward off browsing animals.

Some desert plants produce seed which lies dormant in the soil until sparked into growth by rainfall. When annuals, such plants grow with great rapidity and may flower and set seed within weeks, aiming to complete their development before the last vestige of water dries up. For perennial plants, reproduction is more likely to be successful if the seed germinates in a shaded position, but not so close to the parent plant as to be in competition with it. Some seed will not germinate until it has been blown about on the desert floor to scarify the seed coat. The seed of the mesquite tree which grows in deserts in American is hard and fails to sprout even when planted carefully. When it has passed through the gut of a pronghorn it germinates readily, and the little pile of moist dung provides an excellent start to life well away from the parent tree. The stems and leaves of some plants lower the surface velocity of sand-carrying winds and protect the ground from erosion. Even small fungi and microscopic plant organisms found on the soil surface (so-called *cryptobiotic soil*) can be a vital link in preventing erosion and providing support for other living organisms. Cold deserts often have high concentrations of salt in the soil. Grasses and low shrubs are the dominant vegetation here and the ground may be covered with lichens. Most shrubs have spiny leaves and shed them in the coldest part of the year.

Fauna

Animals adapted to live in deserts are called xerocoles. Many desert animals (and plants) show especially clear evolutionary adaptations for water conservation or heat tolerance and so are often studied in comparative physiology, ecophysiology, and evolutionary physiology. One well-studied example is the specializations of mammalian kidneys shown by desert-inhabiting species. Many examples of convergent evolution have been identified in desert organisms, including between cacti and Euphorbia, kangaroo rats and jerboas, *Phrynosoma* and *Moloch* lizards.

Deserts present a very challenging environment for animals. Not only do they require food and water but they also need to keep their body temperature at a tolerable level. In many ways birds are the most able to do this of the higher animals. They can move to areas of greater food availability as the desert blooms after local rainfall and can fly to faraway waterholes. In hot deserts, gliding birds can remove themselves from the over-heated desert floor by using thermals to soar in the cooler air at great heights. In order to conserve energy, other desert birds run rather than fly. The cream-coloured courser flits gracefully across the ground on its long legs, stopping periodically to snatch up insects. Like other desert birds it is well-camouflaged by its colouring and can merge into the landscape when stationary. The sandgrouse is an expert at this and nests on the open desert floor dozens of kilometres (miles) away from the waterhole it needs to visit daily. Some small diurnal birds are found in very restricted localities where their plumage matches the colour of the underlying surface. The desert lark takes frequent dust baths which ensures that it matches its environment.

Herbivorous mammals obtain moisture from the plants they eat. The addax antelope is so efficient at doing this that it never needs to drink. The camel is a superb example of a mammal adapted to desert life. It minimizes its water loss by producing concentrated urine and dry dung, and is able to lose forty percent of its body weight through water loss without dying of dehydration. Many other hot desert animals are nocturnal, seeking out shade during the day or dwelling underground in burrows. At depths of more than 50 centimetres (20 in), these remain at between 30 and 32 °C (86 and 90 °F) regardless of the external temperature. Jerboas, desert rats, kangaroo rats and other small rodents emerge from their burrows at night and so do the foxes, coyotes, jackals and snakes that prey on them. Carnivores can obtain much of their water needs from the body fluids of their victims.

Being ectotherms, reptiles are unable to live in cold deserts but are well-suited to hot ones. In the heat of the day in the Sahara, the temperature can rise to 50 °C (122 °F). Reptiles cannot survive at this temperature and lizards will be prostrated by heat at 45 °C (113 °F). They have few adaptations to desert life and are unable to cool themselves by sweating so they shelter during the heat of the day. In the first part of the night, as the ground radiates the heat absorbed during the day, they emerge and search for prey. Lizards and snakes are the most numerous in arid regions and certain snakes have developed a novel method of locomotion that enables them to move sidewards and navigate high sand-dunes. These include the horned viper of Africa and the sidewinder of North America, evolutionarily distinct but with similar behavioural patterns because of convergent evolution. Many desert reptiles are ambush predators and often bury themselves in the sand, waiting for a victim to come within range.

Amphibians might seem unlikely desert-dwellers, because of their need to keep their skins moist and their dependence on water for reproductive purposes. In fact, the few species that are found in this habitat have made some remarkable adaptations. Most of them are fossorial, spending the hot dry months aestivating in deep burrows. While there they shed their skins a number of times and retain the remnants around them as a waterproof cocoon to retain moisture. In the Sonoran Desert, Couch's spadefoot toad spends most of the year dormant in its burrow. Heavy rain is the trigger for emergence and the first male to find a suitable pool calls to attract others. Eggs are laid and the tadpoles grow rapidly as they must reach metamorphosis before the water evaporates. As the desert dries out, the adult toads rebury themselves. The juveniles stay on the surface for a while, feeding and growing, but soon dig themselves burrows. Few make it to adulthood. The water holding frog in Australia has a similar life cycle and may aestivate for as long as five years if no rain falls. The Desert rain frog of Namibia is nocturnal and survives because of the damp sea fogs that roll in from the Atlantic.

Invertebrates, particularly arthropods, have successfully made their homes in the desert. Flies, beetles, ants, termites, locusts, millipedes, scorpions and spiders have hard cuticles which are impervious to water and many of them lay their eggs underground and their young develop away from the temperature extremes of the surface. Some make use of the ephemeral pools that form after rain and complete their life cycle in a matter of days. The desert shrimp does this, appearing "miraculously" in new-formed puddles as the dormant eggs hatch. Others, such as brine shrimps, fairy shrimps and tadpole shrimps, are cryptobiotic and can lose up to 92% of their bodyweight, rehydrating as soon as it rains and their temporary pools reappear.

Human Relations

Mineral Extraction: Deserts contain substantial mineral resources, sometimes over their entire surface, giving them their characteristic colours. For example, the red of many sand deserts comes from laterite minerals. Geological processes in a desert climate can concentrate minerals into valuable deposits. Leaching by ground water can extract ore minerals and redeposit them, according to the water table, in concentrated form. Similarly, evaporation tends to concentrate minerals in desert lakes, creating dry lake beds or playas rich in minerals. Evaporation can concentrate minerals as a variety of evaporite deposits, including gypsum, sodium nitrate, sodium chloride and borates. Evaporites are found in the USA's Great Basin Desert, historically exploited by the "20-mule teams" pulling carts of borax from Death Valley to the nearest railway. A desert especially rich in mineral salts is the Atacama Desert, Chile, where sodium nitrate has been mined for explosives and fertilizer since around 1850. Other desert minerals are copper from Chile, Peru, and Iran, and iron and uranium in Australia. Many other metals, salts and commercially valuable types of rock such as pumice are extracted from deserts around the world.

Oil and gas form on the bottom of shallow seas when micro-organisms decompose under anoxic conditions and later become covered with sediment. Many deserts were at one time the sites of shallow seas and others have had underlying hydrocarbon deposits transported to them by the movement of tectonic plates. Some major oilfields such as Ghawar are found under the sands of Saudi Arabia. Geologists believe that other oil deposits were formed by aeolian processes in ancient deserts as may be the case with some of the major American oil fields.

Solar Energy Capture

Deserts are increasingly seen as sources for solar energy, partly due to lower cloud cover. Many successful solar power plants have been built in the Mojave Desert. These plants have a combined capacity of 354 megawatts (MW) making them the largest solar power installation in the world. Large swaths of the desert are covered in mirrors (used for solar energy), including nine fields of solar collectors. The Mojave Solar Park is currently under construction and will produce 280MW when completed.

The potential of generating solar energy from the Sahara desert is immense. Professor David Faiman of Ben-Gurion University has stated that the technology now exists to supply all of the world's electricity needs with 10% of the Sahara desert. Desertec Industrial Initiative is a consortium seeking $560 billion investment in North African solar and wind installations over the next 40 years to supply electricity to Europe via cable lines running under the Mediterranean Sea. European interest in the Sahara desert stems from its two aspects: amount of sunshine and empty space. The Sahara receives more sunshine per are than the sunniest of regions in Europe. The Sahara desert also has the empty space required to house fields of mirrors for solar plants, totalling hundreds of square miles.

The Negev Desert, Israel, and the surrounding area, including the Arava Valley, receive plenty of sunshine and are generally not arable. This has resulted in the construction of many solar plants. David Faiman has proposed that "giant" solar plants in the Negev could supply all of Israel's needs for electricity.

Human Life in Deserts

People have been living in deserts for millennia. Many, such as the Bushmen in the Kalahari, the Aborigines in Australia and various tribes of North American Indians, were originally hunter-gatherers. They developed skills in the manufacture and use of weapons, animal tracking, finding water, foraging for edible plants and using the things they found in their natural environment to supply their everyday needs. Their self-sufficient skills and knowledge were passed down through the generations by word of mouth. Other cultures developed a nomadic way of life as herders of sheep, goats, cattle, camels, yaks, lamas or reindeer. They travelled over large areas with their herds, moving to new pastures as seasonal and erratic rainfall encouraged new plant growth. They took with them their tents made of cloth or skins draped over poles and they ate milk, blood and sometimes meat provided by their livestock.

The desert nomads were also traders. The Sahara is a very large expanse of land stretching from the Atlantic rim to Egypt. Trade routes were developed linking the Sahel in the south with the fertile Mediterranean region to the north and large numbers of camels were used to carry valuable goods across the desert interior. The Tuareg were traders and the goods transported traditionally included slaves, ivory and gold going northwards and salt going southwards. Berbers with knowledge of the region were employed to guide the caravans between the various oases and wells. As many as eight million slaves may have been taken northwards across the Sahara between the 8th and 18th centuries. Traditional means of overland transport declined with the advent of motor vehicles, shipping and air freight, but caravans still travel along routes between Agadez and Bilma and between Timbuktu and Taoudenni carrying salt from the interior to desert-edge communities.

Round the rims of deserts, where more precipitation occurred and conditions were more suitable, some groups took to cultivating crops. This may have happened when drought caused the death of herd animals, forcing herdsmen to turn to cultivation. With few inputs, they were at the mercy of the weather and may have lived at bare subsistence level. The land they cultivated reduced the area available to nomadic herders, causing disputes over land. The semi-arid fringes of the desert have fragile soils which are at risk of erosion when exposed, as happened in the American Dust Bowl in the 1930s. The grasses that held the soil in place were ploughed under, and a series of dry years caused crop failures, while enormous dust storms blew the topsoil away. Half a million Americans were forced to leave their land in this catastrophe.

Similar damage is being done today to the semi-arid areas that rim deserts and about twelve million hectares of land are being turned to desert each year. Desertification is caused by such factors as drought, climatic shifts, tillage for agriculture, overgrazing and deforestation. Vegetation plays a major role in determining the composition of the soil. In many environments, the rate of erosion and run off increases dramatically with reduced vegetation cover. Unprotected dry surfaces tend to be blown away by the wind or be washed away by flash floods, leaving infertile soil layers that bake in the sun and become unproductive hardpan. Although overgrazing has historically been considered to be a cause of desertification, there is some evidence that wild and domesticated animals actually improve fertility and vegetation cover, and that their removal encourages erosive processes.

Grassland

Grasslands are areas where the vegetation is dominated by grasses (Poaceae), however sedge (Cyperaceae) and rush (Juncaceae) families can also be found. Grasslands occur naturally on all continents except Antarctica. In temperate latitudes, such as northwestern Europe and the Great Plains and California in North America, native grasslands are dominated by perennial bunch grass species, whereas in warmer climates annual species form a greater component of the vegetation.

Grasslands are found in most ecoregions of the Earth. For example there are five terrestrial ecoregion classifications (subdivisions) of the temperate grasslands, savannas, and shrublands biome ('ecosystem'), which is one of eight terrestrial ecozones of the Earth's surface.

Vegetation

Grassland vegetation can vary in height from very short, as in chalk where the vegetation may be less than 30 cm (12 in) high, to quite tall, as in the case of North American tallgrass prairie, South American grasslands and African savanna. Woody plants, shrubs or trees, may occur on some grasslands – forming savannas, scrubby grassland or semi-wooded grassland, such as the African savannas or the Iberian dehesa. Such grasslands are sometimes referred to as wood-pasture or woodland. As flowering plants, grasses grow in great concentrations in climates where annual rainfall ranges between 500 and 900 mm (20 and 35 in). The root systems of perennial grasses and forbs form complex mats that hold the soil in place.

Evolution

Graminoids are among the most versatile life forms. They became widespread toward the end of the Cretaceous period, and fossilized dinosaur dung (coprolites) have been found containing phytoliths of a variety of grasses that include grasses that are related to modern rice and bamboo. The appearance of mountains in the western United States during the Miocene and Pliocene epochs, a period of some 25 million years, created a continental climate favourable to the evolution of grasslands. Existing forest biomes declined, and grasslands became much more widespread. Following the Pleistocene Ice Ages, grasslands expanded in range in the hotter, drier climates, and began to become the dominant land feature worldwide.

Climate

Grasslands often occur in areas with annual precipitation between 600 mm (24 in) and 1,500 mm (59 in) and average mean annual temperatures ranges from “5 and 20 °C (Woodward et al. 2004). However, some grasslands occur in colder (20°C) and hotter (30°C) climatic conditions. Grassland can exist in habitats that are frequently disturbed by grazing or fire, as such disturbance prevents the encroachment of woody species. Species richness is particularly high in grasslands of low soil fertility such as serpentine barrens and calcareous grasslands. Infertility may also prevent woody encroachment as low nutrient levels in the soil may inhibit the growth of forest and shrub species.most plants from growing. Temperate grasslands occur in temperate climates typified by distinct seasonality (warm summers and cold winters).

Biodiversity and Conservation

Grasslands dominated by unsown wild-plant communities ("unimproved grasslands") can be called either natural or 'semi-natural' habitats. The majority of grasslands in temperate climates are 'semi-natural'. Although their plant communities are natural, their maintenance depends upon anthropogenic activities such as low-intensity farming, which maintains these grasslands through grazing and cutting regimes. These grasslands contain many species of wild plants – grasses, sedges, rushes and herbs – 25 or more speerican prairie grasslands or lowland wildflower meadows in the UK are now rare and their associated wild flora equally threatened. Associated with the wild-plant diversity of the "unimproved" grasslands is usually a rich invertebrate fauna; also there are many species of birds that are grassland "specialists", such as the snipe and the Great Bustard. Agriculturally improved grasslands, which dominate modern intensive agricultural landscapes, are usually poor in wild plant species due to the original diversity of plants having been destroyed by cultivation, the original wild-plant communities having been replaced by sown monocultures of cultivated varieties of grasses and clovers, such as Perennial ryegrass and White Clover. In many parts of the world "unimproved" grasslands are one of the least threatened habitats, and a target for acquisition by wildlife conservation groups or for special grants to landowners who are encouraged to manage them appropriately.

Human Impact and Economic Importance

Grasslands are of vital importance for raising livestock for human consumption and for milk and other dairy products. Grassland vegetation remains dominant in a particular area usually due to grazing, cutting, or natural or manmade fires, all discouraging colonisation by and survival of tree and shrub seedlings. Some of the world's largest expanses of grassland are found in African savanna, and these are maintained by wild herbivores as well as by nomadic pastoralists and their cattle, sheep or goats. Grasslands may occur naturally or as the result of human activity. Grasslands created and maintained by human activity are called anthropogenic grasslands. Hunting peoples around the world often set regular fires to maintain and extend grasslands, and prevent fire-intolerant trees and shrubs from taking hold. The tallgrass prairies in the US Midwest may have been extended eastward into Illinois, Indiana, and Ohio by human agency. Much grassland in northwest Europe developed after the Neolithic Period, when people gradually cleared the forest to create areas for raising their livestock.

Types

Tropical and Subtropical: These grasslands are classified with tropical and subtropical savannas and shrublands as the tropical and subtropical grasslands, savannas, and shrublands biome. Notable tropical and subtropical grasslands include the Llanos grasslands of northern South America.

Temperate

Mid-latitude grasslands, including the Prairie and Pacific Grasslands of North America, the Pampas of Argentina, Brazil and Uruguay, calcareous downland, and the steppes of Europe. They are classified with temperate savannas and shrublands as the temperate grasslands, savannas, and shrublands biome. Temperate grasslands are the home to many large herbivores, such as bison, gazelles, zebras, rhinoceroses, and wild horses. Carnivores like lions, wolves and cheetahs and leopards are also found in temperate grasslands. Other animals of this region include: deer, prairie dogs, mice, jack rabbits, skunks, coyotes, snakes, fox, owls, badgers, blackbirds (both Old and New World varieties), grasshoppers, meadowlarks, sparrows, quails, hawks and hyenas.

Flooded

Grasslands that are flooded seasonally or year-round, like the Everglades of Florida, the Pantanal of Brazil, Bolivia and Paraguay or the Esteros del Ibera in Argentina.They are classified with flooded savannas as the flooded grasslands and savannas biome and occur mostly in the tropics and subtropics. Watermeadows are grasslands that are deliberately flooded for short periods.

Montane

High-altitude grasslands located on high mountain ranges around the world, like the Páramo of the Andes Mountains. They are part of the montane grasslands and shrublands biome and also constitute tundra.

Tundra

Similar to montane grasslands, polar arctic tundra can have grasses, but high soil moisture means that few tundras are grass-dominated today. However, during the Pleistocene ice ages, a polar grassland known as steppe-tundra occupied large areas of the Northern hemisphere. These are in the tundra biome.

Desert and xeric

Also called desert grasslands, this is composed of sparse grassland ecoregions located in the deserts and xeric shrublands biome.

Fauna

Mites, insect larvae, nematodes and earthworms inhabit deep soil, which can reach 6 metres (20 ft) underground in undisturbed grasslands on the richest soils of the world. These invertebrates, along with symbiotic fungi, extend the root systems, break apart hard soil, enrich it with urea and other natural fertilizers, trap minerals and water and promote growth. Some types of fungi make the plants more resistant to insect and microbial attacks. Grassland in all its form supports a vast variety of mammals, reptiles, birds, and insects. Typical large mammals include the Blue Wildebeest, American Bison, Giant Anteater and Przewalski's Horse. While grasslands in general support diverse wildlife, given the lack of hiding places for predators, the African savanna regions support a much greater diversity in wildlife than do temperate grasslands. There is evidence for grassland being much the product of animal behaviour and movement; some examples include migratory herds of antelope trampling vegetation and African Bush Elephants eating Acacia saplings before the plant has a chance to grow into a mature tree.

Chapter 3

Geographical Distribution of Plants and Animals

Biogeography is the study of the distribution of species and ecosystems in geographic space and through geological time. Organisms and biological communities vary in a highly regular fashion along geographic gradients of latitude, elevation, isolation and habitat area. Knowledge of spatial variation in the numbers and types of organisms is as vital to us today as it was to our early human ancestors, as we adapt to heterogeneous but geographically predictable environments. Biogeography is an integrative field of enquiry that unites concepts and information from ecology, evolutionary biology, geology, and physical geography.

Modern biogeographic research combines information and ideas from many fields, from the physiological and ecological constraints on organismal dispersal to geological and climatological phenomena operating at global spatial scales and evolutionary time frames. The patterns of species distribution across geographical areas can usually be explained through a combination of historical factors such as: speciation; extinction; continental drift; glaciation, and associated variations in sea level, river routes, and habitat; and river capture, in combination with the geographic constraints of landmass areas and isolation; and the available ecosystem energy supplies. Over periods of ecological changes, biogeography includes the study of plant and animal species in: their past and/or present living *refugium* habitat, their interim living sites, and/or their survival locales. As writer David Quammen put it, "...biogeography does more than ask *Which species?* and *Where*. It also asks *Why?* and, what is sometimes more crucial, *Why not?*."

Modern biogeography often employs the use of Geographic Information Systems (GIS), to understand the factors affecting organism distribution, and to predict future trends in organism distribution. Often mathematical models and GIS are employed to solve ecological problems that have a spatial aspect to them.

Biogeography is most keenly observed on the world's islands. These habitats are often much more manageable areas of study because they are more condensed than larger ecosystems on the mainland. Islands are also ideal locations because they allow scientists to look at habitats that new species have only recently colonized and can observe how they disperse throughout the island, the success they achieve in these places, and they can then apply this information to similar mainland habitats. Islands are very diverse in their biomes, ranging from the tropical to arctic climates. This diversity in habitat allows for a wider range of species study in different parts of the world. One scientist who recognized the importance of these geographic locations was Charles Darwin, who remarked in his journal "The Zoology of Archipelagoes will be well worth examination". Two chapters in *On the Origin of Species* were devoted to geographical distribution.

History

The scientific theory of biogeography grows out of the work of Alexander von Humboldt (1769–1859), Hewett Cottrell Watson (1804–1881), Alphonse de Candolle (1806–1893), Alfred Russel Wallace (1823–1913), Philip Lutley Sclater (1829–1913) and other biologists and explorers. Wallace studied the distribution of flora and fauna in the Amazon Basin and the Malay Archipelago in the mid-19th century. Wallace and Sclater saw biogeography as a source of support for the theory of evolution. Key findings, such as the sharp difference in fauna either side of the Wallace Line, and the sharp difference that existed between North and South America prior to their relatively recent faunal interchange, can only be understood in this light. Otherwise, the field of biogeography would be seen as a purely descriptive one. The publication of *The Theory of Island Biogeography* by Robert MacArthur and E.O. Wilson in 1967 showed that the species richness of an area could be predicted in terms of such factors as habitat area, immigration rate and extinction rate. This added to the long-standing interest in island biogeography. The application of island biogeography theory to habitat fragments spurred the development of the fields of conservation biology and landscape ecology.

Classic biogeography has been expanded by the development of molecular systematics, creating a new discipline known as phylogeography. This development allowed scientists to test theories about the origin and dispersal of populations, such as island endemics. For example, while classic biogeographers were able to speculate about the origins of species in the Hawaiian Islands, phylogeography allows them to test theories of relatedness between these populations and putative source populations in Asia and North America.

Paleobiogeography

Paleobiogeography goes one step further to include paleogeographic data and considerations of plate tectonics. Using molecular analyses and corroborated by fossils, it has been possible to demonstrate that perching birds evolved first in the region of Australia or the adjacent Antarctic (which at that time lay somewhat further north and had a temperate climate). From there, they spread to the other Gondwanan continents and Southeast Asia - the part of Laurasia then closest to their origin of dispersal – in the late Paleogene, before achieving a global distribution in the early Neogene. Not knowing the fact that at the time of dispersal, the Indian Ocean was much narrower than it is today, and that South America was closer to the Antarctic, one would be hard pressed to explain the presence of many "ancient" lineages of perching birds in Africa, as well as the mainly South American distribution of the suboscines.

Paleobiogeography also helps constrain hypotheses on the timing of biogeographic events such as vicariance and geodispersal, and provides unique information on the formation of regional biotas. For example, data from species-level phylogenetic and biogeographic studies tell us that the Amazonian fish fauna accumulated incrementally over a period of tens of millions of years, principally by means of allopatric speciation, and in an arena extending over most of the area of tropical South America (Albert & Reis 2011). In other words, unlike some of the well-known insular faunas (Galapagos finches, Hawaiian drosophilid flies, African rift lake cichlids), the species-rich Amazonian ichthyofauna is not the result of recent adaptive radiations. For freshwater organisms, landscapes are divided naturally into discrete drainage basins by watersheds, episodically isolated and reunited by erosional processes. In regions like the Amazon Basin with an exceptionally low (flat) topographic relief, the many waterways have had a highly reticulated history over geological time. In such a context stream capture is an important factor affecting the evolution and distribution of freshwater organisms. Stream capture occurs when an upstream portion of one river drainage is diverted to the downstream portion of an adjacent basin. This can happen as a result of tectonic uplift (or subsidence), natural damming created by a landslide, or headward or lateral erosion of the watershed between adjacent basins.

Classification

Biogeography is a synthetic science, related to geography, biology, soil science, geology, climatology, ecology and evolution. Today, biogeography is broken into three main fields of study. The three fields are historical biogeography, ecological biogeography, and conservation biogeography. Each field, however, looks at phytogeography (the past and present distribution of plants) and zoogeography (the past and present distribution of animals).

Historical biogeography is called paleobiogeography and studies the past distributions of species. It looks at their evolutionary history and things like past climate change to determine why a certain species may have developed in a particular area. For example, the historical approach would say there are more species in the tropics than at high latitudes because the tropics experienced less severe climate change during glacial periods. This led to fewer extinctions and more stable populations over time.

The branch of historical biogeography is called paleobiogeography because it often includes paleogeographic ideas- most notably plate tectonics. This type of research uses fossils to show the movement of species across space via moving continental plates. Paleobiogeography also takes varying climate as a result of the physical land being in different places into account for the presence of different plants and animals. Ecological biogeography looks at the current factors responsible for the distribution of plants and animals. The most common fields of research within ecological biogeography are climatic equability, primary productivity, and habitat heterogeneity. Climatic equability looks at the variation between daily and annual temperatures. It is harder to survive in areas with high variation between day and night and seasonal temperatures. Because of this, there are fewer species at high latitudes because more adaptations are needed to be able to survive there. In contrast, the tropics have a steadier climate with fewer variations in temperature. This means plants do not need to spend their energy on being dormant and then regenerating their leaves and/or flowers, they don't need a flowering season, and they do not need to adapt to extreme hot or cold conditions.

Primary productivity looks at the evapotranspiration rates of plants. Where evapotranspiration is high, so is plant growth. Therefore, areas like the tropics that are warm and moist foster plant transpiration allowing more plants to grow there. In high latitudes, it is simply too cold for the atmosphere to hold enough water vapour to produce high rates of evapotranspiration and there are fewer plants present. Finally, habitat heterogeneity leads to the presence of more biodiversity (a greater number of species present).

After looking at the various fields in historic and ecological biogeography, conservation biogeography developed. This is the protection and/or restoration of nature and its flora and fauna. Biogeography is important as a branch of geography that sheds light on the natural habitats around the world. It is also essential in understanding why species are in their present locations and in developing protecting the world's natural habitats.

Some fundamental concepts in biogeography include:

- evolution – change in genetic composition of a population
- extinction – disappearance of a species
- dispersal – movement of populations away from their point of origin, related to migration
- endemic areas
- geodispersal – the erosion of barriers to biotic dispersal and gene flow, that permit range expansion and the merging of previously isolated biotas
- range and distribution
- vicariance – the formation of barriers to biotic dispersal and gene flow, that tend to subdivide species and biotas, leading to speciation and extinction

Plant

Plants, also called green plants (Viridiplantae in Latin), are living organisms of the kingdom Plantae including such multicellular groups as flowering plants, conifers, ferns and mosses, as well as, depending on definition, the green algae, but not red or brown seaweeds like kelp, nor fungi or bacteria.

Green plants have cell walls with cellulose and characteristically obtain most of their energy from sunlight via photosynthesis using chlorophyll contained in chloroplasts, which gives them their green colour. Some plants are parasitic and may not produce normal amounts of chlorophyll or photosynthesize. Plants are also characterized by sexual reproduction, modular and indeterminate growth, and an alternation of generations, although asexual reproduction is common, and some plants bloom only once while others bear only one bloom.

Precise numbers are difficult to determine, but as of 2010, there are thought to be 300–315 thousand species of plants, of which the great majority, some 260–290 thousand, are seed plants. Green plants provide most of the world's molecular oxygen and are the basis of most of the earth's ecologies, especially on land. Plants described as grains, fruits and vegetables form mankind's basic foodstuffs, and have been domesticated for millennia. Plants serve as ornaments and, until recently and in great variety, they have served as the source of most of medicines and drugs. Their scientific study is known as botany.

Definition

Plants are one of the two groups into which all living things have been traditionally divided; the other is animals. The division goes back at least as far as Aristotle (384 BC – 322 BC) who distinguished between plants which generally do not move, and animals which often are mobile to catch their food. Much later, when Linnaeus (1707–1778) created the basis of the modern system of scientific classification, these two groups became the kingdoms Vegetabilia (later Metaphyta or Plantae) and Animalia (also called Metazoa). Since then, it has become clear that the plant kingdom as originally defined included several unrelated groups, and the fungi and several groups of algae were removed to new kingdoms. However, these organisms are still often considered plants, particularly in popular contexts.

Outside of formal scientific contexts, the term "plant" implies an association with certain traits, such as being multicellular, possessing cellulose, and having the ability to carry out photosynthesis.

Algae

Algae comprise several different groups of organisms which produce energy through photosynthesis and for that reason have been included in the plant kingdom in the past. Most conspicuous among the algae are the seaweeds, multicellular algae that may roughly resemble land plants, but are classified among the brown, red and green algae. Each of these algal groups also includes various microscopic and single-celled organisms. There is good evidence that some of these algal groups arose independently from separate non-photosynthetic ancestors, with the result that many groups of algae are no longer classified within the plant kingdom as it is defined here.

The Viridiplantae, the green plants – green algae and land plants – form a clade, a group consisting of all the descendants of a common ancestor. With a few exceptions among the green algae, all green plants have many features in common, including cell walls containing cellulose, chloroplasts containing chlorophylls *a* and *b*, and food stores in the form of starch. They undergo closed mitosis without centrioles, and typically have mitochondria with flat cristae. The chloroplasts of green plants are surrounded by two membranes, suggesting they originated directly from endosymbiotic cyanobacteria.

Two additional groups, the Rhodophyta (red algae) and Glaucophyta (glaucophyte algae), also have chloroplasts which appear to be derived directly from endosymbiotic cyanobacteria, although they differ in the pigments which are used in photosynthesis and so are different in colour. All three groups together are generally believed to have a single common origin, and so are classified together in the taxon Archaeplastida, whose name implies that the chloroplasts or plastids of all the members of the taxon were derived from a single ancient endosymbiotic event. This is the broadest modern definition of the plants. In contrast, most other algae (e.g. heterokonts, haptophytes, dinoflagellates, and euglenids) not only have different pigments but also have chloroplasts with three or four surrounding membranes. They are not close relatives of the Archaeplastida, presumably having acquired chloroplasts separately from ingested or symbiotic green and red algae. They are thus not included in even the broadest modern definition of the plant kingdom, although they were in the past.

The green plants or Viridiplantae were traditionally divided into the green algae (including the stoneworts) and the land plants. However, it is now known that the land plants evolved from within a group of green algae, so that the green algae by themselves are a paraphyletic group, i.e. a group which excludes some of the descendants of a common ancestor. Paraphyletic groups are generally avoided in modern classifications, so that in recent treatments the Viridiplantae have been divided into two clades, the Chlorophyta and the Streptophyta (or Charophyta). The Chlorophyta (a name that has also been used for *all* green algae) are the sister group to the group from which the land plants evolved. There are about 4,300 species of mainly marine organisms, both unicellular and multicellular. The latter include the sea lettuce, *Ulva*.

The other group within the Viridiplantae are the mainly freshwater or terrestrial Streptophyta (or Charophyta), which consist of several groups of green algae plus the stoneworts and land plants. (The names have been used differently, e.g. Streptophyta to mean the group which excludes the land plants and Charophyta for the stoneworts alone or the stoneworts plus the land plants.) Streptophyte algae are either unicellular or form multicellular filaments, branched or unbranched. The genus *Spirogyra* is a filamentous streptophyte alga familiar to many, as it is often used in teaching and is one of the organisms responsible for the algal "scum" which pond-owners so dislike. The freshwater stoneworts strongly resemble land plants and are believed to be their closest relatives. Growing underwater, they consist of a central stalk with whorls of branchlets, giving them a superficial resemblance to horsetails, species of the genus *Equisetum*, which are true land plants.

Fungi

The classification of fungi has been controversial until quite recently in the history of biology. Linnaeus' original classification placed the fungi within the Plantae, since they were unquestionably not animals or minerals and these were the only other alternatives. With later developments in microbiology, in the 19th century Ernst Haeckel felt that another kingdom was required to classify newly discovered micro-organisms. The introduction of the new kingdom Protista in addition to Plantae and Animalia, led to uncertainty as to whether fungi truly were best placed in the Plantae or whether they ought to be reclassified as protists. Haeckel himself found it difficult to decide and it was not until 1969 that a solution was found whereby Robert Whittaker proposed the creation of the kingdom Fungi. Molecular evidence has since shown that the last common ancestor (concestor) of the Fungi was probably more similar to that of the Animalia than of any other kingdom, including the Plantae.

Whittaker's original reclassification was based on the fundamental difference in nutrition between the Fungi and the Plantae. Unlike plants, which generally gain carbon through photosynthesis, and so are called autotrophic phototrophs, fungi generally obtain carbon by breaking down and absorbing surrounding materials, and so are called heterotrophic saprotrophs. In addition, the substructure of multicellular fungi is different from that of plants, taking the form of many chitinous microscopic strands called hyphae, which may be further subdivided into cells or may form a syncytium containing many eukaryotic nuclei. Fruiting bodies, of which mushrooms are most familiar example, are the reproductive structures of fungi, and are unlike any structures produced by plants.

Evolution

The evolution of plants has resulted in increasing levels of complexity, from the earliest algal mats, through bryophytes, lycopods, ferns to the complex gymnosperms and angiosperms of today. The groups which appeared earlier continue to thrive, especially in the environments in which they evolved.

Evidence suggests that an algal scum formed on the land 1,200 million years ago, but it was not until the Ordovician Period, around 450 million years ago, that land plants appeared. However, new evidence from the study of carbon isotope ratios in Precambrian rocks has suggested that complex photosynthetic plants developed on the earth over 1000 m.y.a. These began to diversify in the late Silurian Period, around 420 million years ago, and the fruits of their diversification are displayed in remarkable detail in an early Devonian fossil assemblage from the Rhynie chert. This chert preserved early plants in cellular detail, petrified in volcanic springs. By the middle of the Devonian Period most of the features recognised in plants today are present, including roots, leaves and secondary wood, and by late Devonian times seeds had evolved. Late Devonian plants had thereby reached a degree of sophistication that allowed them to form forests of tall trees. Evolutionary innovation continued after the Devonian period. Most plant groups were relatively unscathed by the Permo-Triassic extinction event, although the structures of communities changed. This may have set the scene for the evolution of flowering plants in the Triassic (~200 million years ago), which exploded in the Cretaceous and Tertiary. The latest major group of plants to evolve were the grasses, which became important in the mid Tertiary, from around 40 million years ago. The grasses, as well as many other groups, evolved new mechanisms of metabolism to survive the low CO_2 and warm, dry conditions of the tropics over the last 10 million years.

Embryophytes

The plants that are likely most familiar to us are the multicellular land plants, called embryophytes. Embryophytes include the vascular plants, such as ferns, conifers and flowering plants. They also include the *bryophytes*, of which mosses and liverworts are the most common.

All of these plants have eukaryotic cells with cell walls composed of cellulose, and most obtain their energy through photosynthesis, using light, water and carbon dioxide to synthesize food. About three hundred plant species do not photosynthesize but are parasites on other species of photosynthetic plants. Plants are distinguished from green algae, which represent a mode of photosynthetic life similar to the kind modern plants are believed to have evolved from, by having specialized reproductive organs protected by non-reproductive tissues.

Bryophytes first appeared during the early Paleozoic. They can only survive where moisture is available for significant periods, although some species are desiccation tolerant. Most species of bryophytes remain small throughout their life-cycle. This involves an alternation between two generations: a haploid stage, called the gametophyte, and a diploid stage, called the sporophyte. In byrophytes, the sporophyte is always unbranched and typically remains nutritionally dependent on its parent gametophyte.

Vascular plants first appeared during the Silurian period, and by the Devonian had diversified and spread into many different terrestrial environments. They developed a number of adaptations that allowed them to spread into increasingly more arid places. These included a cuticle, a waxy layer that confers resistant to desiccation, stomata, which allow gas exchange while reducing water loss, and vascular tissues, which transport water and food throughout the organism. In vascular plants, the sporophyte is typically large, branched, nutritionally independent and long-lived. Vascular plant gametophytes evolved to become increasingly reduced in size and prominence in the life cycle. The first seed plants, Pteridosperms , now extinct, appeared in the Devonian and diversified through the Carboniferous. In these the microgametophyte is reduced to pollen and the megagametophyte remains inside the megasporangium, attached to the parent plant. A megasporangium invested in protective layer called an integument is known as an ovule. After fertilisation by means of sperm deposited by pollen grains, an embryo develops inside the ovule. The integument becomes a seed coat, and the ovule develops into a seed. Seed plants can survive and reproduce in extremely arid conditions, because they are not dependent on free water for the movement of sperm, or the development of free living gametophytes.

Early seed plants are gymnosperms, as the ovules and subsequent seeds are not enclosed in a protective structure (carpels or fruit), but are found naked, typically on cone scales. Pollen typically lands directly on the ovule. Four surviving groups remain widespread now, particularly the conifers, which are dominant trees in several biomes.

Evolutionary History of plants

The evolution of plants has resulted in increasing levels of complexity, from the earliest algal mats, through bryophytes, lycopods, ferns to the complex gymnosperms and angiosperms of today. While many of the groups which appeared earlier continue to thrive, especially in the environments in which they evolved, for a time each new grade of organisation became more "successful" than its predecessors.

In the Ordovician period, around 450 million years ago, the first land plants appeared. These began to diversify in the late Silurian Period, around 420 million years ago, and the results of their diversification are displayed in remarkable detail in an early Devonian fossil assemblage from the Rhynie chert. This chert preserved early plants in cellular detail, petrified in volcanic springs. By the middle of the Devonian Period, most of the features recognised in plants today are present, including roots, leaves and secondary wood; and, by late Devonian times, seeds had evolved. Late Devonian plants had thereby reached a degree of sophistication that allowed them to form forests of tall trees. Evolutionary innovation continued into the Carboniferous period and is still ongoing today. Most plant groups were relatively unscathed by the Permo-Triassic extinction event, although the structures of communities changed. This may have set the scene for the appearance of the flowering plants in the Triassic (~200 million years ago), and their later diversification in the Cretaceous and Paleogene. The latest major group of plants to evolve were the grasses, which became important in the mid-Paleogene, from around 40 million years ago. The grasses, as well as many other groups, evolved new mechanisms of metabolism to survive the low CO_2 and warm, dry conditions of the tropics over the last 10 million years.

Colonization of Land

Land plants evolved from chlorophyte algae, perhaps as early as 510 million years ago; some molecular estimates place their origin even earlier, as much as 630 million years ago. Their closest living relatives are the charophytes, specifically Charales; assuming that the Charales' habit has changed little since the divergence of lineages, this means that the land plants evolved from a branched, filamentous alga dwelling in shallow fresh water, perhaps at the edge of seasonally desiccating pools. The alga would have had a haplontic life cycle: it would only very briefly have had paired chromosomes (the diploid condition) when the egg and sperm first fused to form a zygote; this would have immediately divided by meiosis to produce cells with half the number of unpaired chromosomes (the haploid condition). Co-operative interactions with fungi may have helped early plants adapt to the stresses of the terrestrial realm.

Plants were not the first photosynthesisers on land; weathering rates suggest that organisms were already living on the land 1,200 million years ago, and microbial fossils have been found in freshwater lake deposits from 1,000 million years ago, but the carbon isotope record suggests that they were too scarce to impact the atmospheric composition until around 850 million years ago. These organisms, although phylogenetically diverse, were probably small and simple, forming little more than an "algal scum".

The first evidence of plants on land comes from spores of Mid-Ordovician age (early Llanvirn, ~470 million years ago). These spores, known as cryptospores, were produced either singly (monads), in pairs (diads) or groups of four (tetrads), and their microstructure resembles that of modern liverwort spores, suggesting they share an equivalent grade of organisation. They are composed of sporopollenin – further evidence of an embryophytic affinity. It could be that atmospheric 'poisoning' prevented eukaryotes from colonising the land prior to this, or it could simply have taken a great time for the necessary complexity to evolve.

Trilete spores similar to those of vascular plants appear soon afterwards, in Upper Ordovician rocks. Depending exactly when the tetrad splits, each of the four spores may bear a "trilete mark", a Y-shape, reflecting the points at which each cell squashed up against its neighbours. However, this requires that the spore walls be sturdy and resistant at an early stage. This resistance is closely associated with having a desiccation-resistant outer wall—a trait only of use when spores must survive out of water. Indeed, even those embryophytes that have returned to the water lack a resistant wall, thus don't bear trilete marks. A close examination of algal spores shows that none have trilete spores, either because their walls are not resistant enough, or in those rare cases where it is, the spores disperse before they are squashed enough to develop the mark, or don't fit into a tetrahedral tetrad.

The earliest megafossils of land plants were thalloid organisms, which dwelt in fluvial wetlands and are found to have covered most of an early Silurian flood plain. They could only survive when the land was waterlogged. There were also microbial mats.

Once plants had reached the land, there were two approaches to dealing with desiccation. The bryophytes avoid it or give in to it, restricting their ranges to moist settings, or drying out and putting their metabolism "on hold" until more water arrives. Tracheophytes resist desiccation: They all bear a waterproof outer cuticle layer wherever they are exposed to air (as do some bryophytes), to reduce water loss, but—since a total covering would cut them off from CO_2 in the atmosphere—they rapidly evolved stomata, small openings to allow gas exchange. Tracheophytes also developed vascular tissue to aid in the movement of water within the organisms , and moved away from a gametophyte dominated life cycle. Vascular tissue also facilitated upright growth without the support of water and paved the way for the evolution of larger plants on land. The establishment of a land-based flora caused increased accumulation of oxygen in the atmosphere, as the plants produced oxygen as a waste product. When this concentration rose above 13%, wildfires became possible. This is first recorded in the early Silurian fossil record by charcoalified plant fossils. Apart from a controversial gap in the Late Devonian, charcoal is present ever since.

Charcoalification is an important taphonomic mode. Wildfire drives off the volatile compounds, leaving only a shell of pure carbon. This is not a viable food source for herbivores or detritovores, so is prone to preservation; it is also robust, so can withstand pressure and display exquisite, sometimes sub-cellular, detail.

Evolution of Life Cycles

All multicellular plants have a life cycle comprising two generations or phases. One is termed the gametophyte, has a single set of chromosomes (denoted 1N), and produces gametes (sperm and eggs). The other is termed the sporophyte, has paired chromosomes (denoted 2N), and produces spores. The gametophyte and sporophyte may appear identical – homomorphy – or may be very different – heteromorphy. The pattern in plant evolution has been a shift from homomorphy to heteromorphy. The algal ancestors of land plants were almost certainly haplobiontic, being haploid for all their life cycles, with a unicellular zygote providing the 2N stage. All land plants (i.e. embryophytes) are diplobiontic – that is, both the haploid and diploid stages are multicellular. Two trends are apparent: bryophytes (liverworts, mosses and hornworts) have developed the gametophyte, with the sporophyte becoming almost entirely dependent on it; vascular plants have developed the sporophyte, with the gametophyte being particularly reduced in the seed plants.

There are two competing theories to explain the appearance of a diplobiontic lifecycle. The interpolation theory (also known as the antithetic or intercalary theory) holds that the sporophyte phase was a fundamentally new invention, caused by the mitotic division of a freshly germinated zygote, continuing until meiosis produces spores. This theory implies that the first sporophytes bore a very different morphology to the gametophyte they depended on. This seems to fit well with what is known of the bryophytes, in which a vegetative thalloid gametophyte is parasitised by simple sporophytes, which often comprise no more than a sporangium on a stalk. Increasing complexity of the ancestrally simple sporophyte, including the eventual acquisition of photosynthetic cells, would free it from its dependence on a gametophyte, as seen in some hornworts (*Anthoceros*), and eventually result in the sporophyte developing organs and vascular tissue, and becoming the dominant phase, as in the tracheophytes (vascular plants). This theory may be supported by observations that smaller *Cooksonia* individuals must have been supported by a gametophyte generation. The observed appearance of larger axial sizes, with room for photosynthetic tissue and thus self-sustainability, provides a possible route for the development of a self-sufficient sporophyte phase.

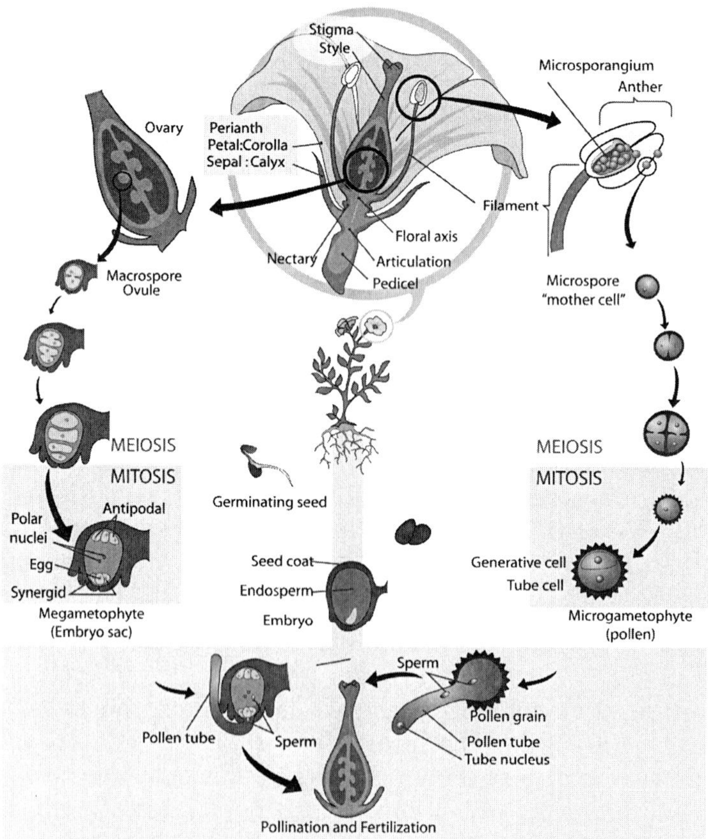

***Figure:** Angiosperm life cycle*

The alternative hypothesis is termed the transformation theory (or homologous theory). This posits that the sporophyte appeared suddenly by a delay in the occurrence of meiosis after the zygote germinated. Since the same genetic material would be employed, the haploid and diploid phases would look the same. This explains the behaviour of some algae, which produce alternating phases of identical sporophytes and gametophytes. Subsequent adaption to the desiccating land environment, which makes sexual reproduction difficult, would result in the simplification of the sexually active gametophyte, and elaboration of the sporophyte phase to better disperse the waterproof spores. The tissue of sporophytes and gametophytes preserved in the Rhynie chert is of similar complexity, which is taken to support this hypothesis.

Evolution of Morphology

Xylem: To photosynthesise, plants must absorb CO_2 from the atmosphere. However, this comes at a price: while stomata are open to allow CO_2 to enter, water can evaporate. Water is lost much faster than CO_2 is absorbed, so plants need to replace it, and have developed systems to transport water from the moist soil to the site of photosynthesis. Early plants sucked water between the walls of their cells, then evolved the ability to control water loss (and CO_2 acquisition) through the use of a waterproof cuticle perforated by stomata. Specialised water transport tissues soon evolved in the form of hydroids, tracheids, then secondary xylem, followed by an endodermis and ultimately vessels.

The high CO_2 levels of Silurian-Devonian times, when plants were first colonising land, meant that the need for water was relatively low. As CO_2 was withdrawn from the atmosphere by plants, more water was lost in its capture, and more elegant transport mechanisms evolved. As water transport mechanisms, and waterproof cuticles, evolved, plants could survive without being continually covered by a film of water. This transition from poikilohydry to homoiohydry opened up new potential for colonisation. Plants then needed a robust internal structure that contained long narrow channels for transporting water from the soil to all the different parts of the above-soil plant, especially to the parts where photosynthesis occurred.

During the Silurian, CO_2 was readily available, so little water needed to be expended to acquire it. By the end of the Carboniferous, when CO_2 levels had lowered to something approaching today's, around 17 times more water was lost per unit of CO_2 uptake. However, even in these "easy" early days, water was at a premium, and had to be transported to parts of the plant from the wet soil to avoid desiccation. This early water transport took advantage of the cohesion-tension mechanism inherent in water. Water has a tendency to diffuse to areas that are drier, and this process is accelerated when water can be wicked along a fabric with small spaces. In small passages, such as that between the plant cell walls (or in tracheids), a column of water behaves like rubber – when molecules evaporate from one end, they literally pull the molecules behind them along the channels. Therefore transpiration alone provided the driving force for water transport in early plants. However, without dedicated transport vessels, the cohesion-tension mechanism cannot transport water more than about 2 cm, severely limiting the size of the earliest plants.

This process demands a steady supply of water from one end, to maintain the chains; to avoid exhausing it, plants developed a waterproof cuticle. Early cuticle may not have had pores but did not cover the entire plant surface, so that gas exchange could continue. However, dehydration at times was inevitable; early plants cope with this by having a lot of water stored between their cell walls, and when it comes to it sticking out the tough times by putting life "on hold" until more water is supplied.

To be free from the constraints of small size and constant moisture that the parenchymatic transport system inflicted, plants needed a more efficient water transport system. During the early Silurian, they developed specialized cells, which were lignified (or bore similar chemical compounds) to avoid implosion; this process coincided with cell death, allowing their innards to be emptied and water to be passed through them. These wider, dead, empty cells were a million times more conductive than the inter-cell method, giving the potential for transport over longer distances, and higher CO_2 diffusion rates.

The earliest macrofossils to bear water-transport tubes are Silurian plants placed in the genus *Cooksonia*. The early Devonian pretracheophytes *Aglaophyton* and *Horneophyton* have structures very similar to the hydroids of modern mosses.

Plants continued to innovate new ways of reducing the resistance to flow within their cells, thereby increasing the efficiency of their water transport. Thickened bands on the walls of tubes are apparent from the early Silurian onwards are adaptations to ease the flow of water. Banded tubes, as well as tubes with pitted ornamentation on their walls, were lignified and, when they form single celled conduits, are referred to as tracheids. These, the "next generation" of transport cell design, have a more rigid structure than hydroids, preventing their collapse at higher levels of water tension. Tracheids may have a single evolutionary origin, possibly within the hornworts, uniting all tracheophytes (but they may have evolved more than once).

Water transport requires regulation, and dynamic control is provided by stomata. By adjusting the amount of gas exchange, they can restrict the amount of water lost through transpiration. This is an important role where water supply is not constant, and indeed stomata appear to have evolved before tracheids, being present in the non-vascular hornworts.

An endodermis probably evolved during the Silu-Devonian, but the first fossil evidence for such a structure is Carboniferous. This structure in the roots covers the water transport tissue and regulates ion exchange (and prevents unwanted pathogens etc. from entering the water transport system). The endodermis can also provide an upwards pressure, forcing water out of the roots when transpiration is not enough of a driver.

Once plants had evolved this level of controlled water transport, they were truly homoiohydric, able to extract water from their environment through root-like organs rather than relying on a film of surface moisture, enabling them to grow to much greater size. As a result of their independence from their surroundings, they lost their ability to survive desiccation – a costly trait to retain.

During the Devonian, maximum xylem diameter increased with time, with the minimum diameter remaining pretty constant. By the mid Devonian, the tracheid diameter of some plant lineages had plateaued. Wider tracheids allow water to be transported faster, but the overall transport rate depends also on the overall cross-sectional area of the xylem bundle itself. The increase in vascular bundle thickness further seems to correlate with the width of plant axes, and plant height; it is also closely related to the appearance of leaves and increased stomatal density, both of which would increase the demand for water.

While wider tracheids with robust walls make it possible to achieve higher water transport pressures, this increases the problem of cavitation. Cavitation occurs when a bubble of air forms within a vessel, breaking the bonds between chains of water molecules and preventing them from pulling more water up with their cohesive tension. A tracheid, once cavitated, cannot have its embolism removed and return to service (except in a few advanced angiosperms that have developed a mechanism of doing so). Therefore, it is well worth plants' while to avoid cavitation occurring. For this reason, pits in tracheid walls have very small diameters, to prevent air entering and allowing bubbles to nucleate. Freeze-thaw cycles are a major cause of cavitation. Damage to a tracheid's wall almost inevitably leads to air leaking in and cavitation, hence the importance of many tracheids working in parallel. Cavitation is hard to avoid, but once it has occurred plants have a range of mechanisms to contain the damage. Small pits link adjacent conduits to allow fluid to flow between them, but not air – although ironically these pits, which prevent the spread of embolisms, are also a major cause of them. These pitted surfaces further reduce the flow of water through the xylem by as much as 30%. Conifers, by the Jurassic, developed an ingenious improvement, using valve-like structures to isolate cavitated elements. These torus-margo structures have a blob floating in the middle of a donut; when one side depressurises the blob is sucked into the torus and blocks further flow. Other plants simply accept cavitation; for instance, oaks grow a ring of wide vessels at the start of each spring, none of which survive the winter frosts. Maples use root pressure each spring to force sap upwards from the roots, squeezing out any air bubbles.

Growing to height also employed another trait of tracheids – the support offered by their lignified walls. Defunct tracheids were retained to form a strong, woody stem, produced in most instances by a secondary xylem. However, in early plants, tracheids were too mechanically vulnerable, and retained a central position, with a layer of tough sclerenchyma on the outer rim of the stems. Even when tracheids do take a structural role, they are supported by sclerenchymatic tissue.

Tracheids end with walls, which impose a great deal of resistance on flow; vessel members have perforated end walls, and are arranged in series to operate as if they were one continuous vessel. The function of end walls, which were the default state in the Devonian, was probably to avoid embolisms. An embolism is where an air bubble is created in a tracheid. This may happen as a result of freezing, or by gases dissolving out of solution. Once an embolism is formed, it usually cannot be removed; the affected cell cannot pull water up, and is rendered useless.

End walls excluded, the tracheids of prevascular plants were able to operate under the same hydraulic conductivity as those of the first vascular plant, *Cooksonia*. The size of tracheids is limited as they comprise a single cell; this limits their length, which in turn limits their maximum useful diameter to 80 ìm. Conductivity grows with the fourth power of diameter, so increased diameter has huge rewards; vessel elements, consisting of a number of cells, joined at their ends, overcame this limit and allowed larger tubes to form, reaching diameters of up to 500 ìm, and lengths of up to 10 m. Vessels first evolved during the dry, low CO_2 periods of the late Permian, in the horsetails, ferns and Selaginellales independently, and later appeared in the mid Cretaceous in angiosperms and gnetophytes. Vessels allow the same cross-sectional area of wood to transport around a hundred times more water than tracheids! This allowed plants to fill more of their stems with structural fibres, and also opened a new niche to vines, which could transport water without being as thick as the tree they grew on. Despite these advantages, tracheid-based wood is a lot lighter, thus cheaper to make, as vessels need to be much more reinforced to avoid cavitation.

Leaves

Leaves today are, in almost all instances, an adaptation to increase the amount of sunlight that can be captured for photosynthesis. Leaves certainly evolved more than once, and probably originated as spiny outgrowths to protect early plants from herbivory.

Leaves are the primary photosynthetic organs of a plant. Based on their structure, they are classified into two types: microphylls, which lack complex venation patterns, and megaphylls, which are large and have complex venation. It has been proposed that these structures arose independently. Megaphylls, according to the Telome hypothesis, have evolved from plants that showed a three dimensional branching architecture, through three transformations—plantation, which involved formation of a planar architecture, webbing, or formation of the outgrowths between the planar branches and fusion, where these webbed outgrowths fused to form a proper leaf lamina. All three steps happened multiple times in the evolution of today's leaves

It has been proposed that before the evolution of leaves, plants had the photosynthetic apparatus on the stems. Today's megaphyll leaves probably became commonplace some 360mya, about 40my after the simple leafless plants had colonized the land in the early Devonian period. This spread has been linked to the fall in the atmospheric carbon dioxide concentrations in the Late Paleozoic era associated with a rise in density of stomata on leaf surface. This must have allowed for better transpiration rates and gas exchange. Large leaves with less stomata would have gotten heated up in the sun's heat, but an increased stomatal density allowed for a better-cooled leaf, thus making its spread feasible.

The rhyniophytes of the Rhynie chert comprised nothing more than slender, unornamented axes. The early to middle Devonian trimerophytes, therefore, are the first evidence of anything that could be considered leafy. This group of vascular plants are recognisable by their masses of terminal sporangia, which adorn the ends of axes which may bifurcate or trifurcate. Some organisms, such as *Psilophyton*, bore enations. These are small, spiny outgrowths of the stem, lacking their own vascular supply.

Around the same time, the zosterophyllophytes were becoming important. This group is recognisable by their kidney-shaped sporangia, which grew on short lateral branches close to the main axes. They sometimes branched in a distinctive H-shape. The majority of this group bore pronounced spines on their axes. However, none of these had a vascular trace, and the first evidence of vascularised enations occurs in the Rhynie genus *Asteroxylon*. The spines of *Asteroxylon* had a primitive vasuclar supply – at the very least, leaf traces could be seen departing from the central protostele towards each individual "leaf". A fossil known as *Baragwanathia* appears in the fossil record slightly earlier, in the late Silurian. In this organism, these leaf traces continue into the leaf to form their mid-vein. One theory, the "enation theory", holds that the leaves developed by outgrowths of the protostele connecting with existing enations, but it is also possible that microphylls evolved by a branching axis forming "webbing".

Asteroxylon and *Baragwanathia* are widely regarded as primitive lycopods. The lycopods are still extant today, familiar as the quillwort *Isoetes* and the club mosses. Lycopods bear distinctive microphylls – leaves with a single vascular trace. Microphylls could grow to some size – the Lepidodendrales boasted microphylls over a meter in length – but almost all just bear the one vascular bundle. (An exception is the branching *Selaginella*).

The more familiar leaves, megaphylls, are thought to have separate origins – indeed, they appeared four times independently, in the ferns, horsetails, progymnosperms, and seed plants. They appear to have originated from dichotomising branches, which first overlapped (or "overtopped") one another, and eventually developed "webbing" and evolved into gradually more leaf-like structures. So megaphylls, by this "teleome theory", are composed of a group of webbed branches – hence the "leaf gap" left where the leaf's vascular bundle leaves that of the main branch resembles two axes splitting. In each of the four groups to evolve megaphylls, their leaves first evolved during the late Devonian to early Carboniferous, diversifying rapidly until the designs settled down in the mid Carboniferous. The cessation of further diversification can be attributed to developmental constraints, but why did it take so long for leaves to evolve in the first place? Plants had been on the land for at least 50 million years before megaphylls became significant. However, small, rare mesophylls are known from the early Devonian genus *Eophyllophyton* – so development could not have been a barrier to their appearance. The best explanation so far incorporates observations that atmospheric CO_2 was declining rapidly during this time – falling by around 90% during the Devonian. This corresponded with an increase in stomatal density by 100 times. Stomata allow water to evaporate from leaves, which causes them to curve. It appears that the low stomatal density in the early Devonian meant that evaporation was limited, and leaves would overheat if they grew to any size. The stomatal density could not increase, as the primitive steles and limited root systems would not be able to supply water quickly enough to match the rate of transpiration.

Clearly, leaves are not always beneficial, as illustrated by the frequent occurrence of secondary loss of leaves, famously exemplified by cacti and the "whisk fern" *Psilotum*.

Secondary evolution can also disguise the true evolutionary origin of some leaves. Some genera of ferns display complex leaves which are attached to the pseudostele by an outgrowth of the vascular bundle, leaving no leaf gap. Further, horsetail (*Equisetum*) leaves bear only a single vein, and appear for all the world to be microphyllous; however, both the fossil record and molecular evidence indicate that their forbears bore leaves with complex venation, and the current state is a result of secondary simplification.

Deciduous trees deal with another disadvantage to having leaves. The popular belief that plants shed their leaves when the days get too short is misguided; evergreens prospered in the Arctic circle during the most recent greenhouse earth. The generally accepted reason for shedding leaves during winter is to cope with the weather – the force of wind and weight of snow are much more comfortably weathered without leaves to increase surface area. Seasonal leaf loss has evolved independently several times and is exhibited in the ginkgoales, pinophyta and angiosperms. Leaf loss may also have arisen as a response to pressure from insects; it may have been less costly to lose leaves entirely during the winter or dry season than to continue investing resources in their repair.

Factors Influencing Leaf Architectures

Various physical and physiological forces like light intensity, humidity, temperature, wind speeds etc. are thought to have influenced evolution of leaf shape and size. It is observed that high trees rarely have large leaves, owing to the obstruction they generate for winds. This obstruction can eventually lead to the tearing of leaves, if they are large. Similarly, trees that grow in temperate or taiga regions have pointed leaves, presumably to prevent nucleation of ice onto the leaf surface and reduce water loss due to transpiration. Herbivory, not only by large mammals, but also small insects has been implicated as a driving force in leaf evolution, an example being plants of the genus Aciphylla, that are commonly found in New Zealand. The now extinct Moas fed upon these plants, and its seen that the leaves have spines on their bodies, which probably functioned to discourage the moas from feeding on them. Other members of Aciphylla, which did not co-exist with the moas, do not have these spines.

At the genetic level, developmental studies have shown that repression of the KNOX genes is required for initiation of the leaf primordium. This is brought about by ARP genes, which encode transcription factors. Genes of this type have been found in many plants studied till now, and the mechanism i.e. repression of KNOX genes in leaf primordia, seems to be quite conserved. Interestingly, expression of KNOX genes in leaves produces complex leaves. It is speculated that the ARP function arose quite early in vascular plant evolution, because members of the primitive group Lycophytes also have a functionally similar gene Other players that have a conserved role in defining leaf primordia are the phytohormone auxin, gibberelin and cytokinin.

One interesting feature of a plant is its phyllotaxy. The arrangement of leaves on the plant body is such that the plant can maximally harvest light under the given constraints, and hence, one might expect the trait to be genetically robust. However, it may not be so. In maize, a mutation in only one gene called ABPHYL (ABnormal PHYLlotaxy) was enough to change the phyllotaxy of the leaves. It implies that sometimes, mutational adjustment of a single locus on the genome is enough to generate diversity. The abphyl gene was later on shown to encode a cytokinin response regulator protein.

Once the leaf primordial cells are established from the SAM cells, the new axes for leaf growth are defined, one important (and more studied) among them being the abaxial-adaxial (lower-upper surface) axes. The genes involved in defining this, and the other axes seem to be more or less conserved among higher plants. Proteins of the HD-ZIPIII family have been implicated in defining the adaxial identity. These proteins deviate some cells in the leaf primordium from the default abaxial state, and make them adaxial. It is believed that, in early plants with leaves, the leaves just had one type of surface — the abaxial one. This is the underside of today's leaves. The definition of the adaxial identity occurred some 200 million years after the abaxial identity was established. One can thus imagine the early leaves as an intermediate stage in evolution of today's leaves, having just arisen from spiny stem-like outgrowths of their leafless ancestors, covered with stomata all over, and not optimized as much for light harvesting.

How the infinite variety of plant leaves is generated is a subject of intense research. Some common themes have emerged. One of the most significant is the involvement of KNOX genes in generating compound leaves, as in the tomato. But, this again is not universal. For example, the pea uses a different mechanism for doing the same thing. Mutations in genes affecting leaf curvature can also change leaf form, by changing the leaf from flat, to a crinky shape, like the shape of cabbage leaves. There also exist different morphogen gradients in a developing leaf which define the leaf's axis. Changes in these morphogen gradients may also affect the leaf form. Another very important class of regulators of leaf development are the microRNAs, whose role in this process has just begun to be documented. The coming years should see a rapid development in comparative studies on leaf development, with many EST sequences involved in the process coming online.

Tree Form

The early Devonian landscape was devoid of vegetation taller than waist height. Without the evolution of a robust vascular system, taller heights could not be attained. There was, however, a constant evolutionary pressure to attain greater height. The most obvious advantage is the harvesting of more sunlight for photosynthesis – by overshadowing competitors – but a further advantage is present in spore distribution, as spores (and, later, seeds) can be blown greater distances if they start higher. This may be demonstrated by *Prototaxites*, thought to be a late Silurian fungus reaching eight metres in height. To attain arborescence, early plants had to develop woody tissue that provided support and water transport. The stele of plants undergoing "secondary growth" is surrounded by the vascular cambium, a ring of cells which produces more xylem (on the inside) and phloem (on the outside). Since xylem cells comprise dead, lignified tissue, subsequent rings of xylem are added to those already present, forming wood. The first plants to develop this secondary growth, and a woody habit, were apparently the ferns, and as early as the middle Devonian one species, *Wattieza*, had already reached heights of 8 m and a tree-like habit.

Other clades did not take long to develop a tree-like stature; the late Devonian *Archaeopteris*, a precursor to gymnosperms which evolved from the trimerophytes, reached 30 m in height. These progymnosperms were the first plants to develop true wood, grown from a bifacial cambium, of which the first appearance is in the mid Devonian *Rellimia*. True wood is only thought to have evolved once, giving rise to the concept of a "lignophyte" clade.

These *Archaeopteris* forests were soon supplemented by lycopods, in the form of lepidodendrales, which topped 50m in height and 2m across at the base. These lycopods rose to dominate late Devonian and Carboniferous coal deposits. Lepidodendrales differ from modern trees in exhibiting determinate growth: after building up a reserve of nutrients at a low height, the plants would "bolt" to a genetically determined height, branch at that level, spread their spores and die. They consisted of "cheap" wood to allow their rapid growth, with at least half of their stems comprising a pith-filled cavity. Their wood was also generated by a unifacial vascular cambium – it did not produce new phloem, meaning that the trunks could not grow wider over time. The horsetail *Calamites* was next on the scene, appearing in the Carboniferous. Unlike the modern horsetail *Equisetum*, *Calamites* had a unifacial vascular cambium, allowing them to develop wood and grow to heights in excess of 10 m. They also branched multiple times.

While the form of early trees was similar to that of today's, the groups containing all modern trees had yet to evolve. The dominant groups today are the gymnosperms, which include the coniferous trees, and the angiosperms, which contain all fruiting and flowering trees. It was long thought that the angiosperms arose from within the gymnosperms, but recent molecular evidence suggests that their living representatives form two distinct groups. The molecular data has yet to be fully reconciled with morphological data, but it is becoming accepted that the morphological support for paraphyly is not especially strong. This would lead to the conclusion that both groups arose from within the pteridosperms, probably as early as the Permian. The angiosperms and their ancestors played a very small role until they diversified during the Cretaceous. They started out as small, damp-loving organisms in the understory, and have been diversifying ever since the mid-Cretaceous, to become the dominant member of non-boreal forests today.

Roots

Roots are important to plants for two main reasons: Firstly, they provide anchorage to the substrate; more importantly, they provide a source of water and nutrients from the soil. Roots allowed plants to grow taller and faster. The onset of roots also had effects on a global scale. By disturbing the soil, and promoting its acidification (by taking up nutrients such as nitrate and phosphate), they enabled it to weather more deeply, promoting the draw-down of CO_2 with huge implications for climate. These effects may have been so profound they led to a mass extinction.

But, how and when did roots evolve in the first place? While there are traces of root-like impressions in fossil soils in the late Silurian, body fossils show the earliest plants to be devoid of roots. Many had tendrils that sprawled along or beneath the ground, with upright axes or thalli dotted here and there, and some even had non-photosynthetic subterranean branches which lacked stomata. The distinction between root and specialised branch is developmental; true roots follow a different developmental trajectory to stems. Further, roots differ in their branching pattern, and in possession of a root cap. So while Silu-Devonian plants such as *Rhynia* and *Horneophyton* possessed the physiological equivalent of roots, roots – defined as organs differentiated from stems – did not arrive until later. Unfortunately, roots are rarely preserved in the fossil record, and our understanding of their evolutionary origin is sparse.

Rhizoids – small structures performing the same role as roots, usually a cell in diameter – probably evolved very early, perhaps even before plants colonised the land; they are recognised in the Characeae, an algal sister group to land plants. That said, rhizoids probably evolved more than once; the rhizines of lichens, for example, perform a similar role. Even some animals (*Lamellibrachia*) have root-like structures! More advanced structures are common in the Rhynie chert, and many other fossils of comparable early Devonian age bear structures that look like, and acted like, roots. The rhyniophytes bore fine rhizoids, and the trimerophytes and herbaceous lycopods of the chert bore root-like structure penetrating a few centimetres into the soil. However, none of these fossils display all the features borne by modern roots. Roots and root-like structures became increasingly more common and deeper penetrating during the Devonian period, with lycopod trees forming roots around 20 cm long during the Eifelian and Givetian. These were joined by progymnosperms, which rooted up to about a metre deep, during the ensuing Frasnian stage. True gymnosperms and zygopterid ferns also formed shallow rooting systems during the Famennian period.

The rhizomorphs of the lycopods provide a slightly different approach to rooting. They were equivalent to stems, with organs equivalent to leaves performing the role of rootlets. A similar construction is observed in the extant lycopod *Isoetes*, and this appears to be evidence that roots evolved independently at least twice, in the lycophytes and other plants, a proposition supported by studies showing that roots are initiated and their growth promoted by different mechanisms in lycophytes and euphyllophytes.

A vascular system is indispensable to rooted plants, as non-photosynthesising roots need a supply of sugars, and a vascular system is required to transport water and nutrients from the roots to the rest of the plant. These plants are little more advanced than their Silurian forbears, without a dedicated root system; however, the flat-lying axes can be clearly seen to have growths similar to the rhizoids of bryophytes today. By the mid-to-late Devonian, most groups of plants had independently developed a rooting system of some nature. As roots became larger, they could support larger trees, and the soil was weathered to a greater depth. This deeper weathering had effects not only on the aforementioned drawdown of CO_2, but also opened up new habitats for colonisation by fungi and animals.

Roots today have developed to the physical limits. They penetrate many metres of soil to tap the water table. The narrowest roots are a mere 40 ìm in diameter, and could not physically transport water if they were any narrower. The earliest fossil roots recovered, by contrast, narrowed from 3 mm to under 700 ìm in diameter; of course, taphonomy is the ultimate control of what thickness can be seen.

Arbuscular Mycorrhizae

The efficiency of many plants' roots is increased via a symbiotic relationship with a fungal partner. The most common are arbuscular mycorrhizae (AM), literally "tree-like fungal roots". These comprise fungi that invade some root cells, filling the cell membrane with their hyphae. They feed on the plant's sugars, but return nutrients generated or extracted from the soil (especially phosphate), which the plant would otherwise have no access to. This symbiosis appears to have evolved early in plant history. AM are found in all plant groups, and 80% of extant vascular plants, suggesting an early ancestry; a "plant"-fungus symbiosis may even have been the step that enabled them to colonise the land,. Such fungi increase the productivity even of simple plants such as liverworts. Indeed, AM are abundant in the Rhynie chert; the association occurred even before there were true roots to colonise, and some have suggested that roots evolved to provide a more comfortable habitat for mycorrhizal fungi.

Factors Affecting Growth

The genotype of a plant affects its growth. For example, selected varieties of wheat grow rapidly, maturing within 110 days, whereas others, in the same environmental conditions, grow more slowly and mature within 155 days. Growth is also determined by environmental factors, such as temperature, available water, available light, and available nutrients in the soil. Any change in the availability of these external conditions will be reflected in the plants growth.

Biotic factors are also capable of affecting plant growth. Plants compete with other plants for space, water, light and nutrients. Plants can be so crowded that no single individual produces normal growth, causing etiolation and chlorosis. Optimal plant growth can be hampered by grazing animals, suboptimal soil composition, lack of mycorrhizal fungi, and attacks by insects or plant diseases, including those caused by bacteria, fungi, viruses, and nematodes.

Simple plants like algae may have short life spans as individuals, but their populations are commonly seasonal. Other plants may be organized according to their seasonal growth pattern: annual plants live and reproduce within one growing season, biennial plants live for two growing seasons and usually reproduce in second year, and perennial plants live for many growing seasons and continue to reproduce once they are mature. These designations often depend on climate and other environmental factors; plants that are annual in alpine or temperate regions can be biennial or perennial in warmer climates. Among the vascular plants, perennials include both evergreens that keep their leaves the entire year, and deciduous plants which lose their leaves for some part of it. In temperate and boreal climates, they generally lose their leaves during the winter; many tropical plants lose their leaves during the dry season.

The growth rate of plants is extremely variable. Some mosses grow less than 0.001 millimeters per hour (mm/h), while most trees grow 0.025-0.250 mm/h. Some climbing species, such as kudzu, which do not need to produce thick supportive tissue, may grow up to 12.5 mm/h. Plants protect themselves from frost and dehydration stress with antifreeze proteins, heat-shock proteins and sugars (sucrose is common). LEA (Late Embryogenesis Abundant) protein expression is induced by stresses and protects other proteins from aggregation as a result of desiccation and freezing.

Plant Cell

Plant cells are typically distinguished by their large water-filled central vacuole, chloroplasts, and rigid cell walls that are made up of cellulose, hemicellulose, and pectin. Cell division is also characterized by the development of a phragmoplast for the construction of a cell plate in the late stages of cytokinesis. Just as in animals, plant cells differentiate and develop into multiple cell types. Totipotent meristematic cells can differentiate into vascular, storage, protective (e.g. epidermal layer), or reproductive tissues, with more primitive plants lacking some tissue types.

Physiology

Photosynthesis: Plants are photosynthetic, which means that they manufacture their own food molecules using energy obtained from light. The primary mechanism plants have for capturing light energy is the pigment chlorophyll. All green plants contain two forms of chlorophyll, chlorophyll *a* and chlorophyll *b*. The latter of these pigments is not found in red or brown algae.

Immune System

By means of cells that behave like nerves, plants receive and distribute within their systems information about incident light intensity and quality. Incident light which stimulates a chemical reaction in one leaf, will cause a chain reaction of signals to the entire plant via a type of cell termed a *bundle sheath cell*. Researchers from the Warsaw University of Life Sciences in Poland, found that plants have a specific memory for varying light conditions which prepares their immune systems against seasonal pathogens. Plants use pattern-recognition receptors to recognize conserved microbial signatures. This recognition triggers an immune response.

The first plant receptors of conserved microbial signatures were identified in rice (XA21, 1995) and in *Arabidopsis thaliana* (FLS2, 2000). Plants also carry immune receptors that recognize highly variable pathogen effectors. These include the NBS-LRR class of proteins.

Internal Distribution

Vascular plants differ from other plants in that they transport nutrients between different parts through specialized structures, called xylem and phloem. They also have roots for taking up water and minerals. The xylem moves water and minerals from the root to the rest of the plant, and the phloem provides the roots with sugars and other nutrient produced by the leaves.

Ecology

The photosynthesis conducted by land plants and algae is the ultimate source of energy and organic material in nearly all ecosystems. Photosynthesis radically changed the composition of the early Earth's atmosphere, which as a result is now 21% oxygen. Animals and most other organisms are aerobic, relying on oxygen; those that do not are confined to relatively rare anaerobic environments. Plants are the primary producers in most terrestrial ecosystems and form the basis of the food web in those ecosystems.

Many animals rely on plants for shelter as well as oxygen and food. Land plants are key components of the water cycle and several other biogeochemical cycles. Some plants have coevolved with nitrogen fixing bacteria, making plants an important part of the nitrogen cycle. Plant roots play an essential role in soil development and prevention of soil erosion.

Distribution

Plants are distributed worldwide in varying numbers. While they inhabit a multitude of biomes and ecoregions, few can be found beyond the tundras at the northernmost regions of continental shelves. At the southern extremes, plants have adapted tenaciously to the prevailing conditions.

Plants are often the dominant physical and structural component of habitats where they occur. Many of the Earth's biomes are named for the type of vegetation because plants are the dominant organisms in those biomes, such as grasslands and forests.

Ecological Relationships

Numerous animals have coevolved with plants. Many animals pollinate flowers in exchange for food in the form of pollen or nectar. Many animals disperse seeds, often by eating fruit and passing the seeds in their feces. Myrmecophytes are plants that have coevolved with ants. The plant provides a home, and sometimes food, for the ants. In exchange, the ants defend the plant from herbivores and sometimes competing plants. Ant wastes provide organic fertilizer.

The majority of plant species have various kinds of fungi associated with their root systems in a kind of mutualistic symbiosis known as mycorrhiza. The fungi help the plants gain water and mineral nutrients from the soil, while the plant gives the fungi carbohydrates manufactured in photosynthesis. Some plants serve as homes for endophytic fungi that protect the plant from herbivores by producing toxins. The fungal endophyte, *Neotyphodium coenophialum*, in tall fescue (*Festuca arundinacea*) does tremendous economic damage to the cattle industry in the U.S. Various forms of parasitism are also fairly common among plants, from the semi-parasitic mistletoe that merely takes some nutrients from its host, but still has photosynthetic leaves, to the fully parasitic broomrape and toothwort that acquire all their nutrients through connections to the roots of other plants, and so have no chlorophyll. Some plants, known as myco-heterotrophs, parasitize mycorrhizal fungi, and hence act as epiparasites on other plants.

Many plants are epiphytes, meaning they grow on other plants, usually trees, without parasitizing them. Epiphytes may indirectly harm their host plant by intercepting mineral nutrients and light that the host would otherwise receive. The weight of large numbers of epiphytes may break tree limbs. Hemiepiphytes like the strangler fig begin as epiphytes but eventually set their own roots and overpower and kill their host. Many orchids, bromeliads, ferns and mosses often grow as epiphytes. Bromeliad epiphytes accumulate water in leaf axils to form phytotelmata, complex aquatic food webs. Approximately 630 plants are carnivorous, such as the Venus Flytrap (*Dionaea muscipula*) and sundew (*Drosera* species). They trap small animals and digest them to obtain mineral nutrients, especially nitrogen and phosphorus.

Importance

The study of plant uses by people is termed economic botany or ethnobotany; some consider economic botany to focus on modern cultivated plants, while ethnobotany focuses on indigenous plants cultivated and used by native peoples. Human cultivation of plants is part of agriculture, which is the basis of human civilization. Plant agriculture is subdivided into agronomy, horticulture and forestry.

Food

Much of human nutrition depends on land plants, either directly or indirectly. Human nutrition depends to a large extent on cereals, especially maize (or corn), wheat and rice. Other staple crops include potato, cassava, and legumes. Human food also includes vegetables, spices, and certain fruits, nuts, herbs, and edible flowers.

Beverages produced from plants include coffee, tea, wine, beer and alcohol. Sugar is obtained mainly from sugar cane and sugar beet. Cooking oils and margarine come from maize, soybean, rapeseed, safflower, sunflower, olive and others. Food additives include gum arabic, guar gum, locust bean gum, starch and pectin. Livestock animals including cows, pigs, sheep, and goats are all herbivores; and feed primarily or entirely on cereal plants, particularly grasses.

Nonfood Products

Wood is used for buildings, furniture, paper, cardboard, musical instruments and sports equipment. Cloth is often made from cotton, flax or synthetic fibres derived from cellulose, such as rayon and acetate. Renewable fuels from plants include firewood, peat and many other biofuels. Coal and petroleum are fossil fuels derived from plants. Medicines derived from plants include aspirin, taxol, morphine, quinine, reserpine, colchicine, digitalis and vincristine.

There are hundreds of herbal supplements such as ginkgo, Echinacea, feverfew, and Saint John's wort. Pesticides derived from plants include nicotine, rotenone, strychnine and pyrethrins. Drugs obtained from plants include opium, cocaine and marijuana. Poisons from plants include ricin, hemlock and curare. Plants are the source of many natural products such as fibres, essential oils, natural dyes, pigments, waxes, tannins, latex, gums, resins, alkaloids, amber and cork. Products derived from plants include soaps, paints, shampoos, perfumes, cosmetics, turpentine, rubber, varnish, lubricants, linoleum, plastics, inks, chewing gum and hemp rope. Plants are also a primary source of basic chemicals for the industrial synthesis of a vast array of organic chemicals. These chemicals are used in a vast variety of studies and experiments.

Aesthetic Uses

Thousands of plant species are cultivated for aesthetic purposes as well as to provide shade, modify temperatures, reduce wind, abate noise, provide privacy, and prevent soil erosion. People use cut flowers, dried flowers and houseplants indoors or in greenhouses. In outdoor gardens, lawn grasses, shade trees, ornamental trees, shrubs, vines, herbaceous perennials and bedding plants are used. Images of plants are often used in art, architecture, humor, language, and photography and on textiles, money, stamps, flags and coats of arms.

Living plant art forms include topiary, bonsai, ikebana and espalier. Ornamental plants have sometimes changed the course of history, as in tulipomania. Plants are the basis of a multi-billion dollar per year tourism industry which includes travel to arboretums, botanical gardens, historic gardens, national parks, tulip festivals, rainforests, forests with colourful autumn leaves and the National Cherry Blossom Festival. Venus Flytrap, sensitive plant and resurrection plant are examples of plants sold as novelties.

Scientific and Cultural Uses

Tree rings are an important method of dating in archeology and serve as a record of past climates. Basic biological research has often been done with plants, such as the pea plants used to derive Gregor Mendel's laws of genetics. Space stations or space colonies may one day rely on plants for life support. Plants are used as national and state emblems, including state trees and state flowers. Ancient trees are revered and many are famous.

Numerous world records are held by plants. Plants are often used as memorials, gifts and to mark special occasions such as births, deaths, weddings and holidays. Plants figure prominently in mythology, religion and literature. The field of ethnobotany studies plant use by indigenous cultures which helps to conserve endangered species as well as discover new medicinal plants. Gardening is the most popular leisure activity in the U.S. Working with plants or horticulture therapy is beneficial for rehabilitating people with disabilities. Certain plants contain psychotropic chemicals which are extracted and ingested, including tobacco, cannabis (marijuana), and opium.

Negative Effects

Weeds are plants that grow where people do not want them. People have spread plants beyond their native ranges and some of these introduced plants become invasive, damaging existing ecosystems by displacing native species. Invasive plants cause billions of dollars in crop losses annually by displacing crop plants, they increase the cost of production and the use of chemical means to control them affects the environment.

Plants may cause harm to animals, including people. Plants that produce windblown pollen invoke allergic reactions in people who suffer from hay fever. A wide variety of plants are poisonous. Toxalbumins are plant poisons fatal to most mammals and act as a serious deterrent to consumption. Several plants cause skin irritations when touched, such as poison ivy. Certain plants contain psychotropic chemicals, which are extracted and ingested or smoked, including tobacco, cannabis (marijuana), cocaine and opium. Smoking causes damage to health or even death, while some drugs may also be harmful or fatal to people. Both illegal and legal drugs derived from plants may have negative effects on the economy, affecting worker productivity and law enforcement costs. Some plants cause allergic reactions when ingested, while other plants cause food intolerances that negatively affect health.

Animal

Animals are multicellular, eukaryotic organisms of the kingdom Animalia or Metazoa. Their body plan eventually becomes fixed as they develop, although some undergo a process of metamorphosis later on in their life. Most animals are motile, meaning they can move spontaneously and independently. All animals must ingest other organisms or their products for sustenance. Most known animal phyla appeared in the fossil record as marine species during the Cambrian explosion, about 542 million years ago. Animals are divided into various sub-groups, including birds, mammals, amphibians, reptiles, fish and insects.

Etymology

The word "animal" comes from the Latin word *animalis*, meaning "having breath". In everyday colloquial usage, the word often refers to non-human members of the kingdom Animalia. Sometimes, only closer relatives of humans such as mammals and other vertebrates are meant in colloquial use. The biological definition of the word refers to all members of the kingdom Animalia, encompassing creatures as diverse as sponges, jellyfish, insects and humans.

Characteristics

Animals have several characteristics that set them apart from other living things. Animals are eukaryotic and multicellular, which separates them from bacteria and most protists. They are heterotrophic, generally digesting food in an internal chamber, which separates them from plants and algae. They are also distinguished from plants, algae, and fungi by lacking rigid cell walls. All animals are motile, if only at certain life stages. In most animals, embryos pass through a blastula stage, which is a characteristic exclusive to animals.

Structure

With a few exceptions, most notably the sponges (Phylum Porifera) and Placozoa, animals have bodies differentiated into separate tissues.

These include muscles, which are able to contract and control locomotion, and nerve tissues, which send and process signals. Typically, there is also an internal digestive chamber, with one or two openings. Animals with this sort of organization are called metazoans, or eumetazoans when the former is used for animals in general.

All animals have eukaryotic cells, surrounded by a characteristic extracellular matrix composed of collagen and elastic glycoproteins. This may be calcified to form structures like shells, bones, and spicules. During development, it forms a relatively flexible framework upon which cells can move about and be reorganized, making complex structures possible. In contrast, other multicellular organisms, like plants and fungi, have cells held in place by cell walls, and so develop by progressive growth. Also, unique to animal cells are the following intercellular junctions: tight junctions, gap junctions, and desmosomes.

Reproduction and Development

Nearly all animals undergo some form of sexual reproduction. They have a few specialized reproductive cells, which undergo meiosis to produce smaller, motile spermatozoa or larger, non-motile ova. These fuse to form zygotes, which develop into new individuals.

Many animals are also capable of asexual reproduction. This may take place through parthenogenesis, where fertile eggs are produced without mating, budding, or fragmentation. A zygote initially develops into a hollow sphere, called a blastula, which undergoes rearrangement and differentiation. In sponges, blastula larvae swim to a new location and develop into a new sponge. In most other groups, the blastula undergoes more complicated rearrangement. It first invaginates to form a gastrula with a digestive chamber, and two separate germ layers — an external ectoderm and an internal endoderm. In most cases, a mesoderm also develops between them. These germ layers then differentiate to form tissues and organs.

Sexual Reproduction Animals

Sexual reproduction is a process that creates a new organism by combining the genetic material of two organisms; it occurs both in eukaryotes and in prokaryotes. A key similarity between bacterial sex and eukaryotic sex is that DNA originating from two different individuals (parents) join up so that homologous sequences are aligned with each other, and this is followed by exchange of genetic information (a process called genetic recombination). After the new recombinant chromosome is formed it is passed on to progeny).

On the other hand, bacterial conjugation, a type of transfer of DNA between two bacteria, is often mistakenly confused with sexual reproduction, because the mechanics are similar. However, bacterial conjugation is controlled by plasmid genes that are adapted for spreading copies of the plasmid between bacteria. The infrequent integration of a plasmid into a host bacterial chromosome, and the subsequent transfer of a part of the host chromosome to another cell do not appear to be bacterial adaptations.

In contrast, bacterial transformation can be regarded as a form of sex in bacteria. Bacterial transformation is a complex process encoded by numerous bacterial genes, and is clearly a bacterial adaptation for DNA transfer. This process occurs naturally in at least 40 bacterial species. For a bacterium to bind, take up, and recombine exogenous DNA into its chromosome, it must enter a special physiological state referred to as competence. Sexual reproduction in early single-celled eukaryotes may have evolved from bacterial transformation.

There are two main processes during sexual reproduction in eukaryotes: meiosis, involving the halving of the number of chromosomes; and fertilization, involving the fusion of two gametes and the restoration of the original number of chromosomes. During meiosis, the chromosomes of each pair usually cross over to achieve homologous recombination.

Sexual reproduction is the primary method of reproduction for the vast majority of macroscopic organisms, including almost all animals and plants. The evolution of sexual reproduction is a major puzzle. The first fossilized evidence of sexual reproduction in organisms such as eukaryotes is in the Stenian period, about 1 to 1.2 billion years ago. Evolutionary thought proposes several explanations for why sexual reproduction developed and why it is maintained.

These reasons include fighting the accumulation of deleterious mutations, increasing rate of adaptation to changing environments, dealing with competition or as an adaptation for repairing DNA damage. The maintenance of sexual reproduction has been explained by theories that work at several different levels of selection, though some of these models remain controversial. New models presented in recent years, however, suggest a basic advantage for sexual reproduction in slowly reproducing, complex organisms, exhibiting characteristics that depend on the specific environment that the given species inhabit, and the particular survival strategies that they employ.

Red Queen hypothesis

The Red Queen hypothesis, also referred to as Red Queen's, Red Queen's race or The Red Queen Effect, is an evolutionary hypothesis which proposes that organisms must constantly adapt, evolve, and proliferate not merely to gain reproductive advantage, but also simply to survive while pitted against ever-evolving opposing organisms in an ever-changing environment. The Red Queen hypothesis intends to explain two different phenomena: the constant extinction rates as observed in the paleontological record caused by co-evolution between competing species and the advantage of sexual reproduction at the level of individuals. The original idea of the Red Queen hypothesis (macroevolutionary) was given by Leigh Van Valen in order to explain the "Law of Extinction". Leigh Van Valen showed that in many populations the probability of extinction does not depend on the lifetime of this population. In addition, the probability of extinction is constant over millions of years for a given population. This could be explained by the coevolution. Indeed, an adaptation in a population of one species (e.g. predators, parasites ...) may change the selection pressure on a population of another species (e.g., prey, hosts), giving rise to an antagonistic coevolution. If this occurs reciprocally, a potential dynamic coevolution may result. The phenomenon's name is derived from a statement that the Red Queen made to Alice in Lewis Carroll's *Through the Looking-Glass* in her explanation of the nature of Wonderland:

Now, *here*, you see, it takes all the running you can do, to keep in the same place.

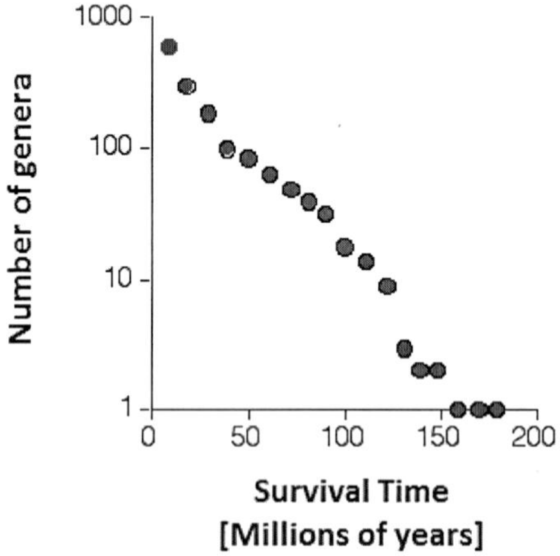

Figure: *The macroevolutionary Red Queen. The linear relationship between number of genera and the logarithm of survival times suggests that the probability of extinction is constant over time. Redrawn from Leigh Van Valen (1973).*

In another idea (microevolutionary), the Red Queen hypothesis is used by Bell to explain the evolution of sex, by John Jaenike to explain the maintenance of sex and W. D. Hamilton explain the role of sex in response to parasite. In all cases, sexual reproduction allows to create variability and a faster response to selection by making offspring genetically unique.

Sexual organisms are able to improve their genotype in function of changing conditions. Consequently a co-evolutionary interactions, between host and parasite for example, may select for a sexual reproduction in hosts in order to reduce the risk of infection. Oscillations in genotype frequencies are observed between parasites and host in an antagonistic coevolution way. Once again, the two stay at the same place.

Tangled Bank Hypotheses

A set of hypotheses, proposed by G. Bell in 1982, which emphasize the importance of the spatial heterogeneity found in a complex environment (i.e. a Tangled Bank) in the maintenance of sexual reproduction. The fitness of a genotype that is successful in one location but less so in another may not be inherited by its offspring. It may be advantageous, therefore, for siblings to disperse at random into microsites nearby, regardless of the local success of the genotype. It is argued that this allows greater penetration of microsites than would be possible for clones of a parthenogenetically reproducing organism and reduces competition among siblings. The 'Tangled Bank' is taken from the opening sentence of the final paragraph of Darwin's Origin of Species: 'It is interesting to contemplate an entangled bank, clothed with many plants of many kinds, with birds singing on the bushes, with various insects flitting about, and with worms crawling through the damp earth, and to reflect that these elaborately constructed forms, so different from each other, and dependent on each other in so complex a manner, have all been produced by laws acting around us.'

Insects

Insect species make up more than two thirds of all extant animal species, and most insect species use sex for reproduction, though some species are facultatively parthenogenetic. Many species have sexual dimorphism, while in others the sexes look nearly identical. Typically they have two sexes with males producing spermatozoa and females ova. The ova develop into eggs that have a covering called the chorion, which forms before internal fertilization. Insects have very diverse mating and reproductive strategies most often resulting in the male depositing spermatophore within the female, which stores the sperm until she is ready for egg fertilization. After fertilization, and the formation of a zygote, and varying degrees of development; the eggs are deposited outside the female in many species, or in some, they develop further within the female and live born offspring are produced.

Reproduction and Development

The majority of insects hatch from eggs, but not all insects lay eggs. The fertilization and development takes place inside the egg, enclosed by a shell (chorion). Some species of insects, like the cockroach *Blaptica dubia*, as well as juvenile aphids and tsetse flies, are ovoviviparous. The eggs of ovoviviparous animals develop entirely inside the female, and then hatch immediately upon being laid. Some other species, such as those in the genus of cockroaches known as *Diploptera*, are viviparous, and thus gestate inside the mother and are born alive. Some insects, like parasitic wasps, show polyembryony, where a single fertilized egg divides into many and in some cases thousands of separate embryos.

Other developmental and reproductive variations include haplodiploidy, polymorphism, paedomorphosis or peramorphosis, sexual dimorphism, parthenogenesis and more rarely hermaphroditism. In haplodiploidy, which is a type of sex-determination system, the offspring's sex is determined by the number of sets of chromosomes an individual receives. This system is typical in bees and wasps. Polymorphism is where a species may have different *morphs* or *forms*, as in the oblong winged katydid, which has four different varieties: green, pink and yellow or tan.

Some insects may retain phenotypes that are normally only seen in juveniles; this is called paedomorphosis. In peramorphosis, an opposite sort of phenomenon, insects take on previously unseen traits after they have matured into adults. Many insects display sexual dimorphism, in which males and females have notably different appearances, such as the moth *Orgyia recens* as an exemplar of sexual dimorphism in insects.

Some insects use parthenogenesis, a process in which the female can reproduce and give birth without having the eggs fertilized by a male. Many aphids undergo a form of parthenogenesis, called cyclical parthenogenesis, in which they alternate between one or many generations of asexual and sexual reproduction. In summer, aphids are generally female and parthenogenetic; in the autumn, males may be produced for sexual reproduction. Other insects produced by parthenogenesis are bees, wasps and ants, in which they spawn males. However, overall, most individuals are female, which are produced by fertilization. The males are haploid and the females are diploid. More rarely, some insects display hermaphroditism, in which a given individual has both male and female reproductive organs.

Insect life-histories show adaptations to withstand cold and dry conditions. Some temperate region insects are capable of activity during winter, while some others migrate to a warmer climate or go into a state of torpor. Still other insects have evolved mechanisms of diapause that allow eggs or pupae to survive these conditions.

Metamorphosis

Metamorphosis in insects is the biological process of development all insects must undergo. There are two forms of metamorphosis: incomplete metamorphosis and complete metamorphosis. Insects that show hemimetabolism, or incomplete metamorphosis, change gradually by undergoing a series of molts. An insect molts when it outgrows its exoskeleton, which does not stretch and would otherwise restrict the insect's growth. The molting process begins as the insect's epidermis secretes a new epicuticle. After this new epicuticle is secreted, the epidermis releases a mixture of enzymes that digests the endocuticle and thus detaches the old cuticle. When this stage is complete, the insect makes its body swell by taking in a large quantity of water or air, which makes the old cuticle split along predefined weaknesses where the old exocuticle was thinnest. Other arthropods have a much different process and only molt; though must accommodate for the difference in exoskeleton structure and make up with other enzymes.

Immature insects that go through incomplete metamorphosis are called nymphs or in the case of dragonflies and damselflies, naiads. Nymphs are similar in form to the adult except for the presence of wings, which are not developed until adulthood. With each molt, nymphs grow larger and become more similar in appearance to adult insects.

Mammals

There are three extant kinds of mammals: Monotremes, Placentals and Marsupials, all with internal fertilization. In placental mammals, offspring are born as juveniles: complete animals with the sex organs present although not reproductively functional. After several months or years, the sex organs develop further to maturity and the animal becomes sexually mature. Most female mammals are only fertile during certain periods during their estrous cycle, at which point they are ready to mate. Individual male and female mammals meet and carry out copulation. For most mammals, males and females exchange sexual partners throughout their adult lives.

Male Placental Mammals

The mammalian male reproductive system contains two main divisions, the penis and the testicles, the latter of which is where sperm are produced. In humans, both of these organs are outside the abdominal cavity, but they can be primarily housed within the abdomen in other animals. For instance, a dog's penis is internal except when mating. Having the testicles outside the abdomen best facilitates temperature regulation of the sperm, which require specific temperatures to survive. The external location may also cause a reduction in the heat-induced contribution to the spontaneous mutation rate in male germinal tissue.

Sperm are the smaller of the two gametes and are generally very short-lived, requiring males to produce them continuously from the time of sexual maturity until death. The produced sperm are stored in the epididymis until ejaculation. The sperm cells are motile and they swim using tail-like flagella to propel themselves towards the ovum. The sperm follows temperature gradients (thermotaxis) and chemical gradients (chemotaxis) to locate the ovum.

Female Placental Mammals

The mammalian female reproductive system likewise contains two main divisions: the vagina and uterus, which act as the receptacle for the sperm, and the ovaries, which produce the female's ova. All of these parts are always internal.

The vagina is attached to the uterus through the cervix, while the uterus is attached to the ovaries via the Fallopian tubes. At certain intervals, the ovaries release an ovum, which passes through the fallopian tube into the uterus.

If, in this transit, it meets with sperm, the egg selects sperm with which to merge; this is termed fertilization. The fertilization usually occurs in the oviducts, but can happen in the uterus itself. The zygote then implants itself in the wall of the uterus, where it begins the processes of embryogenesis and morphogenesis. When developed enough to survive outside the womb, the cervix dilates and contractions of the uterus propel the fetus through the birth canal, which is the vagina.

The ova, which are the female sex cells, are much larger than the sperm and are normally formed within the ovaries of the fetus before its birth. They are mostly fixed in location within the ovary until their transit to the uterus, and contain nutrients for the later zygote and embryo. Over a regular interval, in response to hormonal signals, a process of oogenesis matures one ovum which is released and sent down the Fallopian tube. If not fertilized, this egg is released through menstruation in humans and other great apes, and reabsorbed in other mammals in the estrus cycle.

Gestation

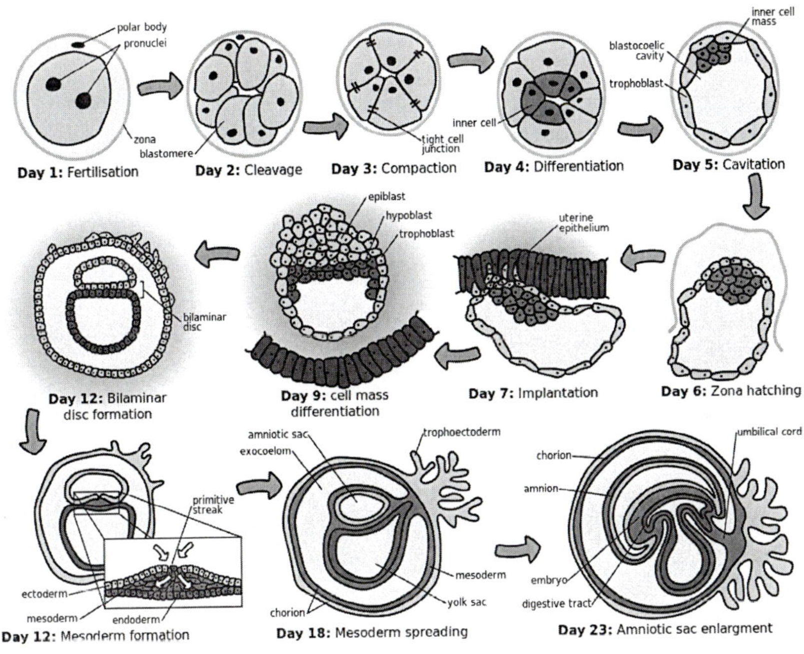

***Figure:** The initial stages of human embryogenesis.*

Gestation, called *pregnancy* in humans, is the period of time during which the fetus develops, dividing via mitosis inside the female. During this time, the fetus receives all of its nutrition and oxygenated blood from the female, filtered through the placenta, which is attached to the fetus' abdomen via an umbilical cord. This drain of nutrients can be quite taxing on the female, who is required to ingest slightly higher levels of calories. In addition, certain vitamins and other nutrients are required in greater quantities than normal, often creating abnormal eating habits. The length of gestation, called the gestation period, varies greatly from species to species; it is 40 weeks in humans, 56–60 in giraffes and 16 days in hamsters.

Birth

Once the fetus is sufficiently developed, chemical signals start the process of birth, which begins with contractions of the uterus and the dilation of the cervix. The fetus then descends to the cervix, where it is pushed out into the vagina, and eventually out of the female. The newborn, which is called an infant in humans, should typically begin respiration on its own shortly after birth. Not long after, the placenta is passed as well. Most mammals eat this, as it is a good source of protein and other vital nutrients needed for caring for the young. The end of the umbilical cord attached to the young's abdomen eventually falls off on its own.

Monotremes

Monotremes, only five species of which exist, all from Australia and New Guinea, are mammals that lay eggs. They have one opening for excretion and reproduction called the cloaca. They hold the eggs internally for several weeks, providing nutrients, and then lay them and cover them like birds. After less than two weeks the young hatches and crawls into its mother's pouch, much like marsupials, where it nurses for several weeks as it grows.

Marsupials

Marsupials' reproductive systems differ markedly from those of placental mammals. The female develops a kind of yolk sac in her womb which delivers nutrients to the embryo. Embryos of some marsupials additionally form placenta-like organs that connect them to the uterine wall, although it is not certain that they transfer nutrients from the mother to the embryo. Pregnancy is very short, typically 4 to 5 weeks, and the embryo is born at a very young stage of development.

Fish

The vast majority of fish species lay eggs that are then fertilized by the male, some species lay their eggs on a substrate like a rock or on plants, while others scatter their eggs and the eggs are fertilized as they drift or sink in the water column. Some fish species use internal fertilization and then disperse the developing eggs or give birth to live offspring. Fish that have live-bearing offspring include the Guppy and Mollies or *Poecilia*. Fishes that give birth to live young can be ovoviviparous, where the eggs are fertilized within the female and the eggs simply hatch within the female body, or in seahorses, the male carries the developing young within a pouch, and gives birth to live young. Fishes can also be viviparous, where the female supplies nourishment to the internally growing offspring. Some fish are hermaphrodites, where a single fish is both male and female and can produce eggs and sperm. In hermaphroditic fish, some are male and female at the same time while in other fish they are serially hermaphroditic; starting as one sex and changing to the other. In at least one hermaphroditic species, self-fertilization occurs when the eggs and sperm are released together. Internal self-fertilization may occur in some other species. One fish species does not reproduce by sexual reproduction but uses sex to produce offspring; *Poecilia formosa* is a unisex species that uses a form of parthenogenesis called gynogenesis, where unfertilized eggs develop into embryos that produce female offspring. *Poecilia formosa* mate with males of other fish species that use internal fertilization, the sperm does not fertilize the eggs but stimulates the growth of the eggs which develops into embryos.

Asexual Reproduction Examples in Animals

Asexual reproduction is a mode of reproduction by which offspring arise from a single parent, and inherit the genes of that parent only; it is reproduction which almost always does not involve meiosis, ploidy reduction, or fertilization. The offspring will be exact genetic copies of the parent, except in the specific case of automixis. A more stringent definition is agamogenesis which is reproduction without the fusion of gametes. Asexual reproduction is the primary form of reproduction for single-celled organisms such as the archaea, bacteria, and protists. Many plants and fungi reproduce asexually as well.

While all prokaryotes reproduce asexually (without the formation and fusion of gametes), mechanisms for lateral gene transfer such as conjugation, transformation and transduction are sometimes likened to sexual reproduction. A complete lack of sexual reproduction is relatively rare among multicellular organisms, particularly animals. It is not entirely understood why the ability to reproduce sexually is so common among them. Current hypotheses suggest that asexual reproduction may have short term benefits when rapid population growth is important or in stable environments, while sexual reproduction offers a net advantage by allowing more rapid generation of genetic diversity, allowing adaptation to changing environments. Developmental constraints may underlie why few animals have relinquished sexual reproduction completely in their life-cycles. Another constraint on switching from sexual to asexual reproduction would be the concomitant loss of meiosis and the protective recombinational repair of DNA damage afforded as one function of meiosis.

Types of Asexual Reproduction

Fission: In binary fission the parent organism is replaced by two daughter organisms, because it literally divides in two. Organisms, both prokaryotes (the archaea and the bacteria), and eukaryotes (such as protists and unicellular fungi), reproduce asexually through binary fission; most of these are also capable of sexual reproduction. In multiple fission, the nucleus of the parent cell divides by mitosis several times, producing several nuclei. The cytoplasm then separates, creating multiple daughter cells.

Budding

Some cells split via budding (for example baker's yeast), resulting in a 'mother' and 'daughter' cell. The offspring organism is smaller than the parent. Budding is also known on a multicellular level; an animal example is the hydra, which reproduces by budding. The buds grow into fully matured individuals which eventually break away from the parent organism. For example, if potatoes are left in warm conditions for too long they begin to sprout from the buds, which are commonly referred to as eyes.

Vegetative Propagation

Vegetative reproduction is a type of asexual reproduction found in plants where new individuals are formed without the production of seeds or spores by meiosis or syngamy. Examples of vegetative reproduction include the formation of miniaturized plants called plantlets on specialized leaves (for example in kalanchoe) and some produce new plants out of rhizomes or stolon (for example in strawberry). Other plants reproduce by forming bulbs or tubers (for example tulip bulbs and dahlia tubers). Some plants produce adventitious shoots and suckers that form along their lateral roots. Plants that reproduce vegetatively may form a clonal colony, where all the individuals are clones, and the clones may cover a large area.

Spore Formation

Many multicellular organisms form spores during their biological life cycle in a process called *sporogenesis*. Exceptions are animals and some protists, who undergo *gametic meiosis* immediately followed by fertilization. Plants and many algae on the other hand undergo *sporic meiosis* where meiosis leads to the formation of haploid spores rather than gametes. These spores grow into multicellular individuals (called gametophytes in the case of plants) without a fertilization event. These haploid individuals give rise to gametes through mitosis. Meiosis and gamete formation therefore occur in separate generations or "phases" of the life cycle, referred to as alternation of generations. Since sexual reproduction is often more narrowly defined as the fusion of gametes (fertilization), spore formation in plant sporophytes and algae might be considered a form of asexual reproduction (agamogenesis) despite being the result of meiosis and undergoing a reduction in ploidy. However, both events (spore formation and fertilization) are necessary to complete sexual reproduction in the plant life cycle.

Fungi and some algae can also utilize true asexual spore formation, which involves mitosis giving rise to reproductive cells called mitospores that develop into a new organism after dispersal. This method of reproduction is found for example in conidial fungi and the red alga *Polysiphonia*, and involves sporogenesis without meiosis. Thus the chromosome number of the spore cell is the same as that of the parent producing the spores. However, mitotic sporogenesis is an exception and most spores, such as those of plants, most Basidiomycota, and many algae, are produced by meiosis.

Fragmentation

Fragmentation is a form of asexual reproduction where a new organism grows from a fragment of the parent. Each fragment develops into a mature, fully grown individual. Fragmentation is seen in many organisms such as animals (some annelid worms, turbellarians and sea stars), fungi, and plants. Some plants have specialized structures for reproduction via fragmentation, such as *gemma* in liverworts. Most lichens, which are a symbiotic union of a fungus and photosynthetic algae or bacteria, reproduce through fragmentation to ensure that new individuals contain both symbiont. These fragments can take the form of *soredia*, dust-like particles consisting of fungal hyphen wrapped around photobiont cells.

Agamogenesis

Agamogenesis is any form of reproduction that does not involve a male gamete. Examples are parthenogenesis and apomixis.

Parthenogenesis

Parthenogenesis is a form of agamogenesis in which an unfertilized egg develops into a new individual. Parthenogenesis occurs naturally in many plants, invertebrates (e.g. water fleas, rotifers, aphids, stick insects, some ants, bees and parasitic wasps), and vertebrates (e.g. some reptiles, amphibians, rarely birds). In plants, apomixis may or may not involve parthenogenesis.

Apomixis and Nucellar Embryony

Apomixis in plants is the formation of a new sporophyte without fertilization. It is important in ferns and in flowering plants, but is very rare in other seed plants. In flowering plants, the term "apomixis" is now most often used for agamospermy, the formation of seeds without fertilization, but was once used to include vegetative reproduction. An example of an apomictic plant would be the triploid European dandelion. Apomixis mainly occurs in two forms: In gametophytic apomixis, the embryo arises from an unfertilized egg within a diploid embryo sac that was formed without completing meiosis. In nucellar embryony, the embryo is formed from the diploid nucellus tissue surrounding the embryo sac. Nucellar embryony occurs in some citrus seeds. Male apomixis can occur in rare cases, such as the Saharan Cypress Cupressus dupreziana, where the genetic material of the embryo are derived entirely from pollen. The term "apomixis" is also used for asexual reproduction in some animals, notably water-fleas, Daphnia.

Alternation between Sexual and Asexual Reproduction

Some species alternate between the sexual and asexual strategies, an ability known as heterogamy, depending on conditions. Alternation is observed in several rotifer species and a few types of insects, such as aphids which will, under certain conditions, produce eggs that have not gone through meiosis, thus cloning themselves. The cape bee *Apis mellifera* subsp. *capensis* can reproduce asexually through a process called thelytoky. A few species of amphibians, reptiles, and birds have a similar ability. For example, the freshwater crustacean *Daphnia* reproduces by parthenogenesis in the spring to rapidly populate ponds, then switches to sexual reproduction as the intensity of competition and predation increases. Another example are monogonont rotifers of the genus Brachionus, which reproduce via cyclical parthenogenesis: at low population densities females produce asexually and at higher densities a chemical cue accumulates and induces the transition to sexual reproduction. Many protists and fungi alternate between sexual and asexual reproduction.

For example, the slime mold *Dictyostelium* undergoes binary fission (mitosis) as single-celled amoebae under favourable conditions. However, when conditions turn unfavourable, the cells aggregate and follow one of two different developmental pathways, depending on conditions. In the social pathway, they form a multicellular slug which then forms a fruiting body with asexually generated spores. In the sexual pathway, two cells fuse to form a giant cell that develops into a large cyst. When this macrocyst germinates, it releases hundreds of amoebic cells that are the product of meiotic recombination between the original two cells.

The hyphae of the common mold (*Rhizopus*) are capable of producing both mitotic as well as meiotic spores. Many algae similarly switch between sexual and asexual reproduction. A number of plants use both sexual and asexual means to produce new plants, some species alter their primary modes of reproduction from sexual to asexual under varying environmental conditions.

Inheritance of Asexual Reproduction in Sexual Species

For example, in the rotifer *Brachionus calyciflorus* asexual reproduction (obligate parthenogenesis) can be inherited by a recessive allele, which leads to loss of sexual reproduction in homozygous offspring. Inheritance of asexual reproduction by a single recessive locus has also been found in the parasitoid wasp *Lysiphlebus fabarum*.

Examples in Animals

There are examples of parthenogenesis in the hammerhead shark and the blacktip shark. In both cases, the sharks had reached sexual maturity in captivity in the absence of males, and in both cases the offspring were shown to be genetically identical to the mothers.

Reptiles use the ZW sex-determination system, which produces either males (with ZZ sex chromosomes) or females (with ZW or WW sex chromosomes). Until 2010, it was thought that the ZW chromosome system used by reptiles was incapable of producing viable WW offspring, but a (ZW) female boa constrictor was discovered to have produced viable female offspring with WW chromosomes. The female boa could have chosen any number of male partners (and had successfully in the past) but on these occasions she reproduced asexually, creating 22 female babies with WW sex-chromosomes.

Polyembryony is a widespread form of asexual reproduction in animals, whereby the fertilized egg or a later stage of embryonic development splits to form genetically identical clones. While in animals, this phenomenon has been best studied in the parasitic Hymenoptera. In the 9-banded armadillos, this process is obligatory and usually gives rise to genetically identical quadruplets. In other mammals, monozygotic twinning has no apparent genetic basis, though its occurrence is common. There are at least 10 million identical human twins and triplets in the world today. Bdelloid rotifers reproduce exclusively asexually, and all individuals in the class Bdelloidea are females. Asexuality evolved in these animals millions of years ago and has persisted since. There is evidence to suggest that asexual reproduction has allowed the animals to evolve new proteins through the Meselson effect that have allowed them to survive better in periods of dehydration. Molecular evidence strongly suggest that at least two species of the stick insect genus *Timema* have used only asexual (parthenogenetic) reproduction for one million years, the longest period known for any insect.

Food and Energy Sourcing

All animals are heterotrophs, meaning that they feed directly or indirectly on other living things. They are often further subdivided into groups such as carnivores, herbivores, omnivores, and parasites.

Predation is a biological interaction where a predator (a heterotroph that is hunting) feeds on its prey (the organism that is attacked). Predators may or may not kill their prey prior to feeding on them, but the act of predation always results in the death of the prey. The other main category of consumption is detritivory, the consumption of dead organic matter. It can at times be difficult to separate the two feeding behaviours, for example, where parasitic species prey on a host organism and then lay their eggs on it for their offspring to feed on its decaying corpse. Selective pressures imposed on one another has led to an evolutionary arms race between prey and predator, resulting in various antipredator adaptations.

Most animals indirectly use the energy of sunlight by eating plants or plant-eating animals. Most plants use light to convert inorganic molecules in their environment into carbohydrates, fats, proteins and other biomolecules, characteristically containing reduced carbon in the form of carbon-hydrogen bonds. Starting with carbon dioxide (CO_2) and water (H_2O), photosynthesis converts the energy of sunlight into chemical energy in the form of simple sugars (e.g., glucose), with the release of molecular oxygen. These sugars are then used as the building blocks for plant growth, including the production of other biomolecules. When an animal eats plants (or eats other animals which have eaten plants), the reduced carbon compounds in the food become a source of energy and building materials for the animal. They are either used directly to help the animal grow, or broken down, releasing stored solar energy, and giving the animal the energy required for motion. Animals living close to hydrothermal vents and cold seeps on the ocean floor are not dependent on the energy of sunlight. Instead chemosynthetic archaea and bacteria form the base of the food chain.

Origin and Fossil Record

Animals are generally considered to have evolved from a flagellated eukaryote. Their closest known living relatives are the choanoflagellates, collared flagellates that have a morphology similar to the choanocytes of certain sponges. Molecular studies place animals in a supergroup called the opisthokonts, which also include the choanoflagellates, fungi and a few small parasitic protists. The name comes from the posterior location of the flagellum in motile cells, such as most animal spermatozoa, whereas other eukaryotes tend to have anterior flagella. The first fossils that might represent animals appear in the Trezona Formation at Trezona Bore, West Central Flinders, South Australia. These fossils are interpreted as being early sponges. They were found in 665-million-year-old rock.

The next oldest possible animal fossils are found towards the end of the Precambrian, around 610 million years ago, and are known as the Ediacaran or Vendian biota. These are difficult to relate to later fossils, however. Some may represent precursors of modern phyla, but they may be separate groups, and it is possible they are not really animals at all. Aside from them, most known animal phyla make a more or less simultaneous appearance during the Cambrian period, about 542 million years ago. It is still disputed whether this event, called the Cambrian explosion, represents a rapid divergence between different groups or a change in conditions that made fossilization possible.

Some paleontologists suggest that animals appeared much earlier than the Cambrian explosion, possibly as early as 1 billion years ago. Trace fossils such as tracks and burrows found in the Tonian era indicate the presence of triploblastic worms, like metazoans, roughly as large (about 5 mm wide) and complex as earthworms. During the beginning of the Tonian period around 1 billion years ago, there was a decrease in Stromatolite diversity, which may indicate the appearance of grazing animals, since stromatolite diversity increased when grazing animals went extinct at the End Permian and End Ordovician extinction events, and decreased shortly after the grazer populations recovered. However the discovery that tracks very similar to these early trace fossils are produced today by the giant single-celled protist *Gromia sphaerica* casts doubt on their interpretation as evidence of early animal evolution. It has been estimated that 99.9% of animals that have ever existed are extinct.

Urmetazoan

The Urmetazoan is the hypothetical last common ancestor of all animals. It was undoubtedly marine, but beyond this its form is difficult to determine; five different hypotheses have been put forward, and discriminating between them is difficult because the relationships of animal phyla are not completely resolved.

Animals are considered by biologists to have evolved from a flagellated eukaryote. Their closest known living relatives are the choanoflagellates, collared flagellates that have a morphology similar to the choanocytes of certain sponges. Molecular studies place animals in a supergroup called the opisthokonts, which also include the choanoflagellates, fungi and a few small parasitic protists. The name comes from the posterior location of the flagellum in motile cells, such as most animal spermatozoa, whereas other eukaryotes tend to have anterior flagella.

Placula Hypothesis

Under the placula hypothesis, the last common ancestor of animals was a somewhat amorphous blob, lacking any form of symmetry or axis. The centre of this blob became raised slightly from the sediment, to create a cavity which assisted feeding on the sea floor below. As this cavity developed, it became deeper and deeper, until the organisms resembled a thimble, with an inside and outside. Some sponges and cnidaria have this kind of body form.

The development of the bilaterian body plan follows from this hypothesis; the urbilaterian would develop its symmetry as one end of the placula became adapted for moving forwards, which would result in a left-right symmetry.

Planula Hypothesis

This idea suggests that metazoa are derived from a planula; that is, the tadpole-like larva of certain cnidaria. Under this hypothesis, the larva became sexually mature through paedomorphosis, so could reproduce without passing through a sessile phase.

Gastraea Hypothesis

The gastraea hypothesis was proposed by Ernst Haeckel, shortly after his work on the calcareous sponges. He proposed that this group of sponges is monophyletic with all eumetazoans, including the bilaterians. This suggested that the gastrulation and the gastrula stage are universal for eumetazoans. It has been perceived as problematic that a gastrulation by invagination is by no means universal among eumetazoans. Only recently has an invagination been confirmed in a Calcarea sponge, albeit too early to form a remaining inner space (archenteron).

Bilaterogastraea Hypothesis

This hypothesis was developed by Gösta Jägersten as an adaptation of Ernst Haeckel's Gastraea hypothesis. He proposed that the Bilaterogastraea had a two-stage life cycle, with a pelagic juvenile and a benthic adult stage. The invagination of the original gastrula stage he saw as bilaterally symmetric rather than radially symmetric.

Groups of Animals

Porifera, Radiata and basal Bilateria

Phylogenetic analysis suggests that the Porifera and Ctenophora diverged before a clade that gave rise to the Bilateria, Cnidaria and Placozoa. Another study based on the presence/absence of introns suggests that Cnidaria, Porifera and Placozoa may be a sister group of Bilateria and Ctenophora.

The sponges (Porifera) were long thought to have diverged from other animals early. They lack the complex organization found in most other phyla. Their cells are differentiated, but in most cases not organized into distinct tissues. Sponges typically feed by drawing in water through pores. Archaeocyatha, which have fused skeletons, may represent sponges or a separate phylum. However, a phylogenomic study in 2008 of 150 genes in 29 animals across 21 phyla revealed that it is the Ctenophora or comb jellies which are the basal lineage of animals, at least among those 21 phyla. The authors speculate that sponges—or at least those lines of sponges they investigated—are not so primitive, but may instead be secondarily simplified.

Among the other phyla, the Ctenophora and the Cnidaria, which includes sea anemones, corals, and jellyfish, are radially symmetric and have digestive chambers with a single opening, which serves as both the mouth and the anus. Both have distinct tissues, but they are not organized into organs. There are only two main germ layers, the ectoderm and endoderm, with only scattered cells between them. As such, these animals are sometimes called diploblastic. The tiny placozoans are similar, but they do not have a permanent digestive chamber.

The remaining animals form a monophyletic group called the Bilateria. For the most part, they are bilaterally symmetric, and often have a specialized head with feeding and sensory organs. The body is triploblastic, i.e. all three germ layers are well-developed, and tissues form distinct organs. The digestive chamber has two openings, a mouth and an anus, and there is also an internal body cavity called a coelom or pseudocoelom. There are exceptions to each of these characteristics, however — for instance adult echinoderms are radially symmetric, and certain parasitic worms have extremely simplified body structures.

Genetic studies have considerably changed our understanding of the relationships within the Bilateria. Most appear to belong to two major lineages: the deuterostomes and the protostomes, the latter of which includes the Ecdysozoa, Platyzoa, and Lophotrochozoa. In addition, there are a few small groups of bilaterians with relatively similar structure that appear to have diverged before these major groups. These include the Acoelomorpha, Rhombozoa, and Orthonectida. The Myxozoa, single-celled parasites that were originally considered Protozoa, are now believed to have developed from the Medusozoa as well.

Deuterostomes

Deuterostomes differ from the other Bilateria, called protostomes, in several ways. In both cases there is a complete digestive tract. However, in protostomes, the first opening of the gut to appear in embryological development (the archenteron) develops into the mouth, with the anus forming secondarily. In deuterostomes the anus forms first, with the mouth developing secondarily. In most protostomes, cells simply fill in the interior of the gastrula to form the mesoderm, called schizocoelous development, but in deuterostomes, it forms through invagination of the endoderm, called enterocoelic pouching. Deuterostome embryos undergo radial cleavage during cell division, while protostomes undergo spiral cleavage. All this suggests the deuterostomes and protostomes are separate, monophyletic lineages. The main phyla of deuterostomes are the Echinodermata and Chordata. The former are radially symmetric and exclusively marine, such as starfish, sea urchins, and sea cucumbers. The latter are dominated by the vertebrates, animals with backbones. These include fish, amphibians, reptiles, birds, and mammals. In addition to these, the deuterostomes also include the Hemichordata, or acorn worms. Although they are not especially prominent today, the important fossil graptolites may belong to this group. The Chaetognatha or arrow worms may also be deuterostomes, but more recent studies suggest protostome affinities.

Ecdysozoa

The Ecdysozoa are protostomes, named after the common trait of growth by moulting or ecdysis. The largest animal phylum belongs here, the Arthropoda, including insects, spiders, crabs, and their kin. All these organisms have a body divided into repeating segments, typically with paired appendages. Two smaller phyla, the Onychophora and Tardigrada, are close relatives of the arthropods and share these traits. The ecdysozoans also include the Nematoda or roundworms, perhaps the second largest animal phylum. Roundworms are typically microscopic, and occur in nearly every environment where there is water. A number are important parasites. Smaller phyla related to them are the Nematomorpha or horsehair worms, and the Kinorhyncha, Priapulida, and Loricifera. These groups have a reduced coelom, called a pseudocoelom. The remaining two groups of protostomes are sometimes grouped together as the Spiralia, since in both embryos develop with spiral cleavage.

Platyzoa

The Platyzoa include the phylum Platyhelminthes, the flatworms. These were originally considered some of the most primitive Bilateria, but it now appears they developed from more complex ancestors. A number of parasites are included in this group, such as the flukes and tapeworms. Flatworms are acoelomates, lacking a body cavity, as are their closest relatives, the microscopic Gastrotricha.

The other platyzoan phyla are mostly microscopic and pseudocoelomate. The most prominent are the Rotifera or rotifers, which are common in aqueous environments. They also include the Acanthocephala or spiny-headed worms, the Gnathostomulida, Micrognathozoa, and possibly the Cycliophora. These groups share the presence of complex jaws, from which they are called the Gnathifera.

Lophotrochozoa

The Lophotrochozoa include two of the most successful animal phyla, the Mollusca and Annelida. The former, which is the second largest animal phylum by number of described species, includes animals such as snails, clams, and squids, and the latter comprises the segmented worms, such as earthworms and leeches. These two groups have long been considered close relatives because of the common presence of trochophore larvae, but the annelids were considered closer to the arthropods because they are both segmented. Now, this is generally considered convergent evolution, owing to many morphological and genetic differences between the two phyla. The Lophotrochozoa also include the Nemertea or ribbon worms, the Sipuncula, and several phyla that have a ring of ciliated tentacles around the mouth, called a lophophore. These were traditionally grouped together as the lophophorates. But it now appears that the lophophorate group may be paraphyletic, with some closer to the nemerteans and some to the molluscs and annelids. They include the Brachiopoda or lamp shells, which are prominent in the fossil record, the Entoprocta, the Phoronida, and possibly the Bryozoa or moss animals.

Model Organisms

Because of the great diversity found in animals, it is more economical for scientists to study a small number of chosen species so that connections can be drawn from their work and conclusions extrapolated about how animals function in general. Because they are easy to keep and breed, the fruit fly *Drosophila melanogaster* and the nematode *Caenorhabditis elegans* have long been the most intensively studied metazoan model organisms, and were among the first life-forms to be genetically sequenced. This was facilitated by the severely reduced state of their genomes, but as many genes, introns, and linkages lost, these ecdysozoans can teach us little about the origins of animals in general. The extent of this type of evolution within the superphylum will be revealed by the crustacean, annelid, and molluscan genome projects currently in progress. Analysis of the starlet sea anemone genome has emphasised the importance of sponges, placozoans, and choanoflagellates, also being sequenced, in explaining the arrival of 1500 ancestral genes unique to the Eumetazoa.

An analysis of the homoscleromorph sponge *Oscarella carmela* also suggests that the last common ancestor of sponges and the eumetazoan animals was more complex than previously assumed.

Other model organisms belonging to the animal kingdom include the house mouse (*Mus musculus*) and zebrafish (*Danio rerio*).

History of Classification

Aristotle divided the living world between animals and plants, and this was followed by Carolus Linnaeus (Carl von Linné), in the first hierarchical classification. Since then biologists have begun emphasizing evolutionary relationships, and so these groups have been restricted somewhat. For instance, microscopic protozoa were originally considered animals because they move, but are now treated separately. In Linnaeus's original scheme, the animals were one of three kingdoms, divided into the classes of Vermes, Insecta, Pisces,

Amphibia, Aves, and Mammalia. Since then the last four have all been subsumed into a single phylum, the Chordata, whereas the various other forms have been separated out. The above lists represent our current understanding of the group, though there is some variation from source to source.

Fauna

Fauna is all of the animal life of any particular region or time. The corresponding term for plants is *flora*. Flora, fauna and other forms of life such as fungi are collectively referred to as biota.

Zoologists and paleontologists use *fauna* to refer to a typical collection of animals found in a specific time or place, e.g. the "Sonoran Desert fauna" or the "Burgess Shale fauna".

Paleontologists sometimes refer to a sequence of faunal stages, which is a series of rocks all containing similar fossils.

"Fauna" comes from the Latin names of Fauna, a Roman goddess of earth and fertility, the Roman god Faunus, and the related forest spirits called Fauns. All three words are cognates of the name of the Greek god Pan, and *panis* is the Greek equivalent of fauna. *Fauna* is also the word for a book that catalogues the animals in such a manner. The term was first used by Linnaeus in the title of his 1745 work *Fauna Suecica*.

Subdivisions of Fauna

Cryofauna: *Cryofauna* are animals that live in, or very close to, ice.

Cryptofauna: *Cryptofauna* are animals that are rarely seen and may be extinct or mythological.

Infauna: *Infauna* are benthic organisms that live within the bottom substratum of a body of water, especially within the bottom-most oceanic sediments, rather than on its surface. Bacteria and microalgae may also live in the interstices of bottom sediments. In general infaunal animals become progressively smaller and less abundant with increasing water depth and distance from shore, whereas bacteria show more constancy in abundance, tending toward one billion cells per milliliter of interstitial seawater.

Epifauna

Epifauna, also called *epibenthos*, are aquatic animals that live on the bottom substratum as opposed to within it, that is, the benthic fauna that live on top of the sediment surface at the seafloor.

Macrofauna

Macrofauna are benthic or soil organisms which are retained on a 0.5mm sieve. Studies in the deep sea define macrofauna as animals retained on a 0.3mm sieve to account for the small size of many of the taxa.

Megafauna

Megafauna are large animals of any particular region or time. For example, Australian megafauna. In terrestrial zoology, megafauna (Ancient Greek *megas* "large" + New Latin *fauna* "animal") are large or giant animals. The most common thresholds used are 45 kilograms (100 lb) or 100 kilograms (220 lb). This thus includes many species not popularly thought of as overly large, such as white-tailed deer and red kangaroo, and even humans.

In practice, the most common usage encountered in academic and popular writing describes land animals roughly larger than a human that are not (solely) domesticated. The term is especially associated with the Pleistocene megafauna — the land animals often larger than modern counterparts considered archetypical of the last ice age, such as mammoths, the majority of which in northern Eurasia, the Americas and Australia became extinct as recently as 10,000-40,000 years ago. It is also commonly used for the largest extant wild land animals, especially elephants, giraffes, hippopotamuses, rhinoceroses, and large bovines. Megafauna may be subcategorized by their trophic position into megaherbivores (e.g. elk), megacarnivores (e.g. lions), and more rarely, megaomnivores (e.g. bears). Other common uses are for giant aquatic species, especially whales, any larger wild or domesticated land animals such as larger antelope and cattle, and dinosaurs and other extinct giant reptilians. The term is also sometimes applied to animals (usually extinct) of great size *relative to* a more common or surviving type of the animal, for example the 1 m (3 ft) dragonflies of the Carboniferous period.

Ecological Strategy

Megafauna — in the sense of the largest mammals and birds — are generally K-strategists, with high longevity, slow population growth rates, low mortality rates, and (at least for the largest) few or no natural predators capable of killing adults. These characteristics, although not exclusive to such megafauna, make them vulnerable to human overexploitation, in part because of their slow population recovery rates.

Evolution of Large Body Size

One observation that has been made about the evolution of larger body size is that rapid rates of increase that are often seen over relatively short time intervals are not sustainable over much longer time periods. In an examination of mammal body mass changes over time, the maximum increase possible in a given time interval was found to scale with the interval length raised to the 0.25 power. This is thought to reflect the emergence, during a trend of increasing maximum body size, of a series of anatomical, physiological, environmental, genetic and other constraints that must be overcome by evolutionary innovations before further size increases are possible. A strikingly faster rate of change was found for large decreases in body mass, such as may be associated with the phenomenon of insular dwarfism. When normalized to generation length, the maximum rate of body mass decrease was found to be over 30 times greater than the maximum rate of body mass increase for a ten-fold change.

In Terrestrial Mammals

Subsequent to the Cretaceous–Paleogene extinction event that eliminated the dinosaurs about 66 Ma ago, terrestrial mammals underwent a nearly exponential increase in body size as they diversified to occupy the ecological niches left vacant. Starting from just a few kg before the event, maximum size had reached ~50 kg a few million years later, and ~750 kg by the end of the Paleocene. This trend of increasing body mass appears to level off about 40 Ma ago (in the late Eocene), suggesting that physiological or ecological constraints had been reached, after an increase in body mass of over three orders of magnitude. However, when considered from the standpoint of rate of size increase per generation, the exponential increase is found to have continued until the appearance of *Indricotherium* 30 Ma ago. (Since generation time scales with *body mass*$^{0.295}$, increasing generation times with increasing size cause the log mass vs. time plot to curve downward from a linear fit.)

Megaherbivores eventually attained a body mass of over 10,000 kg. The largest of these, indricotheres and proboscids, have been hindgut fermenters, which are believed to have an advantage over foregut fermenters in terms of being able to accelerate gastrointestinal transit in order to accommodate very large food intakes. A similar trend emerges when rates of increase of maximum body mass per generation for different mammalian clades are compared (using rates averaged over macroevolutionary time scales). Among terrestrial mammals, the fastest rates of increase of *body mass*$^{0.259}$ vs. time (in Ma) occurred in perissodactyls (a slope of 2.1), followed by rodents (1.2) and proboscids (1.1), all of which are hindgut fermenters. The rate of increase for artiodactyls (0.74) was about a third that of perissodactyls. The rate for carnivorans (0.65) was slightly lower yet, while primates, perhaps constrained by their arboreal habits, had the lowest rate (0.39) among the mammalian groups studied.

Terrestrial mammalian carnivores from several eutherian groups (the mesonychid *Andrewsarchus*, the creodonts *Megistotherium* and *Sarkastodon*, and the carnivorans *Amphicyon* and *Arctodus*) all reached a maximum size of about 1000 kg (the carnivoran *Arctotherium* apparently actually got somewhat larger). The largest known metatherian carnivore, *Proborhyaena gigantea*, apparently reached 600 kg, also close to this limit. A similar theoretical maximum size for mammalian carnivores has been predicted based on the metabolic rate of mammals, the energetic cost of obtaining prey, and the maximum estimated rate coefficient of prey intake. It has also been suggested that maximum size for mammalian carnivores is constrained by the stress the humerus can withstand at top running speed.

Analysis of the variation of maximum body size over the last 40 Ma suggests that decreasing temperature and increasing continental land area are associated with increasing maximum body size. The former correlation would be consistent with Bergmann's rule, and might be related to the thermoregulatory advantage of large body mass in cool climates, better ability of larger organisms to cope with seasonality in food supply, or other factors; the latter correlation could be explainable in terms of range and resource limitations. However, the two parameters are interrelated (due to sea level drops accompanying increased glaciation), making the driver of the trends in maximum size more difficult to identify.

In Marine Mammals

The ancestors of cetaceans are believed to have been the semiaquatic pakicetids, no larger than wolves, of about 53 million years (Ma) ago. By 40 Ma ago, cetaceans had attained a length of 20 m or more in *Basilosaurus*, an elongated, serpentine whale that differed from modern whales in many respects and was not ancestral to them. Following this, the evolution of large body size in cetaceans appears to have come to a temporary halt, and then to have backtracked, although the available fossil records are limited. However, in the period from 31 Ma ago (in the Oligocene) to the present, cetaceans underwent a significantly more rapid sustained increase in body mass (a rate of increase in *body mass*$^{0.295}$ of a factor of 3.2 per million years) than achieved by any group of terrestrial mammals. This trend led to the largest animal of all time, the modern blue whale. Several reasons for the more rapid evolution of large body size in cetaceans are possible. Fewer biomechanical constraints on increases in body size may be associated with suspension in water as opposed to standing against the force of gravity, and with swimming movements as opposed to terrestrial locomotion. Also, the greater heat capacity and thermal conductivity of water compared to air may increase the thermoregulatory advantage of large body size in marine endotherms, although diminishing returns apply.

Cetaceans are not the only marine mammals to reach unprecedented size in the modern era. The largest carnivoran of all time is the mostly aquatic modern southern elephant seal.

In Flightless Birds

During the Paleocene, because of the small initial size of all mammals, apex predator niches were often occupied by members of other classes, such as terrestrial crocodilians (e.g. *Pristichampsus*), large snakes (e.g. *Titanoboa*), varanid lizards, or flightless birds (e.g. *Gastornis* in Europe and North America, *Paleopsilopterus* in South America). In the northern continents, large predatory birds were displaced when large eutherian carnivores evolved. In isolated South America, the phorusrhacids could not be outcompeted by the local metatherian sparassodonts and remained dominant until advanced eutherian predators arrived from North America (as part of the Great American Interchange) during the Pliocene. However, none of the largest predatory (*Brontornis*), possibly omnivorous (*Dromornis*) or herbivorous (*Aepyornis*) flightless birds of the Cenozoic ever grew to masses much above 500 kg, and thus never attained the size of the largest mammalian carnivores, let alone that of the largest mammalian herbivores. It has been suggested that the increasing thickness of avian eggshells in proportion to egg mass with increasing egg size places an upper limit on the size of birds. The largest species of *Dromornis*, *D. stirtoni*, may have gone extinct after it attained the maximum avian body mass and was then outcompeted by marsupial diprotodonts that evolved to sizes several times larger

Megafaunal Mass Extinctions

Timing and Possible Causes: A well-known mass extinction of megafauna, the Holocene extinction, occurred at the end of the last ice age glacial period (a.k.a. the Würm glaciation) and wiped out many giant ice age animals, such as woolly mammoths, in the Americas and northern Eurasia. Various theories have attributed the wave of extinctions to human hunting, climate change, disease, a putative extraterrestrial impact, or other causes. However, this extinction pulse near the end of the Pleistocene was just one of a series of megafaunal extinction pulses that have occurred during the last 50,000 years over much of the Earth's surface, with Africa and southern Asia being largely spared. The latter areas did suffer a gradual attrition of megafauna, particularly of the slower-moving species (a class of vulnerable megafauna epitomized by giant tortoises), over the last several million years.

Outside the mainland of Afro-Eurasia, these megafaunal extinctions followed a distinctive landmass-by-landmass pattern that closely parallels the spread of humans into previously uninhabited regions of the world, and which shows no correlation with climatic history (which can be visualized with plots over recent geological time periods of climate markers such as marine oxygen isotopes or atmospheric carbon dioxide levels). Australia was struck first around 45,000 years ago, followed by Tasmania about 41,000 years ago (after formation of a land bridge to Australia about 43,000 years ago), Japan apparently about 30,000 years ago, North America 13,000 years ago, South America about 500 years later, Cyprus 10,000 years ago, the Antilles 6000 years ago, New Caledonia and nearby islands 3000 years ago, Madagascar 2000 years ago, New Zealand 700 years ago, the Mascarenes 400 years ago, and the Commander Islands 250 years ago. Nearly all of the world's isolated islands could furnish similar examples of extinctions occurring shortly after the arrival of *Homo sapiens*, though most of these islands, such as the Hawaiian Islands, never had terrestrial megafauna, so their extinct fauna were smaller.

An analysis of *Sporormiella* fungal spores (which derive mainly from the dung of megaherbivores) in swamp sediment cores spanning the last 130,000 years from Lynch's Crater in Queensland, Australia showed that the megafauna of that region virtually disappeared about 41,000 years ago, at a time when climate changes were minimal; the change was accompanied by an increase in charcoal, and was followed by a transition from rainforest to fire-tolerant sclerophyll vegetation.

The high-resolution chronology of the changes supports the hypothesis that human hunting alone eliminated the megafauna, and that the subsequent change in flora was most likely a consequence of the elimination of browsers and an increase in fire. The increase in fire lagged the disappearance of megafauna by about a century, and most likely resulted from accumulation of fuel once browsing stopped. Over the next several centuries grass increased; sclerophyll vegetation increased with a lag of another century, and a sclerophyll forest developed after about another thousand years.

During two periods of climate change about 120 and 75 thousand years ago, sclerophyll vegetation had also increased at the site in response to a shift to cooler, drier conditions; neither of these episodes had a significant impact on megafaunal abundance.

Continuing human hunting and environmental disturbance has led to additional megafaunal extinctions in the recent past, and has created a serious danger of further extinctions in the near future.

A number of other mass extinctions occurred earlier in Earth's geologic history, in which some or all of the megafauna of the time also died out. Famously, in the Cretaceous–Paleogene extinction event the dinosaurs and most other giant reptilians were eliminated. However, the earlier mass extinctions were more global and not so selective for megafauna; i.e., many species of other types, including plants, marine invertebrates and plankton, went extinct as well. Thus, the earlier events must have been caused by more generalized types of disturbances to the biosphere.

Effect on Methane Emissions

Large populations of megaherbivores have the potential to contribute greatly to the atmospheric concentration of methane, which is an important greenhouse gas. Modern ruminant herbivores produce methane as a byproduct of foregut fermentation in digestion, and release it through belching. Today, around 20% of annual methane emissions come from livestock methane release. In the Mesozoic, it has been estimated that sauropods could have emitted 520 million tons of methane to the atmosphere annually, contributing to the warmer climate of the time (up to 10 °C warmer than at present). This large emission follows from the enormous estimated biomass of sauropods, and because methane production of individual herbivores is believed to be almost proportional to their mass.

Recent studies have indicated that the extinction of megafaunal herbivores may have caused a reduction in atmospheric methane. This hypothesis is relatively new. One study examined the methane emissions from the bison that occupied the Great Plains of North

America before contact with European settlers. The study estimated that the removal of the bison caused a decrease of as much as 2.2 million tons per year. Another study examined the change in the methane concentration in the atmosphere at the end of the Pleistocene epoch after the extinction of megafauna in the Americas. After early humans migrated to the Americas about 13,000 BP, their hunting and other associated ecological impacts led to the extinction of many megafaunal species there. Calculations suggest that this extinction decreased methane production by about 9.6 milion tons per year. This suggests that the absence of megafaunal methane emissions may have contributed to the abrupt climatic cooling at the onset of the Younger Dryas. The decrease in atmospheric methane that occurred at that time, as recorded in ice cores, was 2-4 times more rapid than any other decrease in the last half million years, suggesting that an unusual mechanism was at work.

Meiofauna

Meiofauna are small benthic invertebrates that live in both marine and fresh water environments. The term *Meiofauna* loosely defines a group of organisms by their size, larger than microfauna but smaller than macrofauna, rather than a taxonomic grouping. One environment for meiofauna is between grains of damp sand. In practice these are metazoan animals that can pass unharmed through a 0.5 – 1 mm mesh but will be retained by a 30 – 45 ìm mesh, but the exact dimensions will vary from researcher to researcher. Whether an organism passes through a 1 mm mesh also depends upon whether it is alive or dead at the time of sorting.

Mystacocarida

Mystacocarida is a subclass of crustaceans, that form part of the meiobenthos. They are less than 1 mm (0.04 in) long, and live interstitially in the intertidal zones of sandy beaches.

Taxonomy

The taxonomy of Mystacocarida is extremely conservative, since all mystacocarids look superficially alike. As of October 2012, there are 13 described species divided between two genera, *Derocheilocaris* (8 species) and *Ctenocheilocharis* (5 species). The first mystacocarids to be found were discovered on a beach on southern New England in 1939.

Decsription

Mystacocarids are tiny crustaceans, less than 1 millimetre (0.04 in) long, that live in the spaces between sand grains on intertidal beaches.

They have a cylindrical body, with five thoracic and five abdominal segments. There are four pairs of small thoracic appendages. The head is relatively large and divided into two by a stricture, so that the latter part gives the appearance of being a part of the thorax. This region bears a pair of maxillipeds, and the head also has two pairs of maxillae, a pair of limb-like mandibles, and two pairs of long antennae. The appendages on the head are much longer than those on the thorax, and have a number of fine hairs that the animal uses to strain detritus from the water to feed on. They have a single naupliar eye. After mating, mystacocarids lay tiny eggs which hatch into a nauplius or metanauplius larva.

Mesofauna

Mesofauna are macroscopic soil invertebrates such as arthropods or nematodes. Mesofauna are extremely diverse; considering just the springtails (Collembola), as of 1998, approximately 6,500 species had been identified.

Microfauna

Microfauna are microscopic or very small animals (usually including protozoans and very small animals such as rotifers). Microfauna (Ancient Greek *mikros* "small" + New Latin *fauna* "animal") refers to microscopic organisms that exhibit animal-like qualities. Microfauna are represented in the animal kingdom (e.g., nematodes, small arthropods) and the protist kingdom (i.e., protozoans). This is in contrast to microflora which, together with microfauna, make up the microzoa.

Habitat

Microfauna are present in every habitat on Earth. They fill essential roles as decomposers and food sources for lower trophic levels, and are necessary to drive processes within larger organisms.

Role

One particular example of the role of microfauna can be seen in soil, where they are important in the cycling of nutrients in ecosystems. Soil microfauna are capable of digesting just about any organic substance, and some inorganic substances (such as TNT and synthetic rubber). These organisms are often essential links in the food chain between primary producers and larger species. For example, zooplankton are widespread microscopic animals and protists which feed on algae and detritus in the ocean. They include foraminifora and krill, which are the primary food source for even animals such as whales. Microfauna also aid in digestion and other processes in larger organisms.

Cryptozoa

The microfauna are the least understood of soil life, due to their small size and great diversity. Many microfauna are members of the so-called cryptozoa, animals that remain undescribed by science. Out of the estimated 10-20 million animal species in the world, only 1.8 million have been given scientific names, and many of the remaining millions are likely microfauna, much of it from the tropics.

Other

Other terms include *avifauna*, which means "bird fauna" and *piscifauna* (or *ichthyofauna*), which means "fish fauna".

Chapter 4

Ecosystem Conservation and Management

Ecosystem management is a process that aims to conserve major ecological services and restore natural resources while meeting the socioeconomic, political and cultural needs of current and future generations. The principal objective of ecosystem management is the efficient maintenance, and ethical use of natural resources. Ecosystem management acknowledges that the interrelation of socio-cultural, economic and ecological systems is paramount to understanding the circumstances that affect environmental goals and outcomes. It is a multifaceted and holistic approach which requires a significant change in how the natural and human environments are identified. Several approaches to effective ecosystem management engage conservation efforts at both a local or landscape level and involves: adaptive management, natural resource management, strategic management, and command and control management.

Formulations

The definitions of ecosystem management are typically vague. Several core principles define and bound the concept and provide operational meaning: (1) ecosystem management reflects a stage in the continuing evolution of social values and priorities, it is neither a beginning nor an end; (2) ecosystem management is place-based and the boundaries of the place must be clearly and formally defined; (3) ecosystem management should maintain ecosystems in the appropriate condition to achieve desired social benefits; (4) ecosystem management should take advantage of the ability of ecosystems to respond to a variety of stressors, natural and man-made, but all ecosystems have limited ability to accommodate stressors and maintain a desired state;

(5) ecosystem management may or may not result in emphasis on biological diversity; (6) the term sustainability, if used at all in ecosystem management, should be clearly defined—specifically, the time frame of concern, the benefits and costs of concern, and the relative priority of the benefits and costs; and (7) scientific information is important for effective ecosystem management, but is only one element in a decision-making process that is fundamentally one of public and private choice. As a concept of natural resource management, ecosystem management remains both ambiguous and controversial, in part because some of its formulations rest on policy and scientific assertions that are contested. These assertions are important to understanding much of the conflict surrounding ecosystem management. Professional natural resource managers, typically operating from within government bureaucracies and professional organizations, often mask debate over controversial assertions by depicting ecosystem management as an evolution of past management approaches.

Stakeholders

Stakeholders are individuals or groups of people who are affected by environmental decisions and actions, but they also may have power to influence the outcomes of environmental decisions relating to ecosystem management (Reed et al. 2009). The complex nature of decisions made in ecosystem management, from local to international scales, requires stakeholder participation from a diversity of knowledge, perceptions and values of nature (Billgren and Holmen 2007; and Reed 2008). Stakeholders will often have different interests in ecosystem services (Shepherd 2008) which means effective management of ecosystems requires a negotiation process that develops mutual trust in issues of common interest with the objective of creating mutually beneficial partnerships (Mushove and Vogel 2005).

Adaptive Management

Adaptive management is based on the concept that predicting future influences/disturbance to an ecosystem is limited and unclear (Pahl-Wostl 2007). Therefore, the goal of adaptive management is to manage the ecosystem so it maintains the greatest amount of ecological integrity, but also to utilize management practices that have the ability to change based on new experience and insights (Department of Interior, Holling 1978, Pahl-Wostl 2007).

Adaptive management aims to identify uncertainties in the management of an ecosystem while using hypothesis testing to further understand the system (Resilience Alliance 2010). In this regard, adaptive management encourages learning from the outcomes of previously implemented management strategies (Pahl-Wostl 2007). Ecosystem managers form hypotheses about the ecosystem and its functionality and then implement different management techniques to test the hypotheses (DOI). The implemented techniques are then analyzed to evaluate any regressions or improvements in functionality of the ecosystem caused by the technique (DOI). Further analysis allows for modification of the technique until it successfully meets the ecological needs of the ecosystem (Holling 1978). Thus, adaptive management serves as a "learning by doing" method for ecosystem management.

Adaptive management has had mixed success in the field of ecosystem management (Gregory et al. 2006). This is because ecosystem managers may not be equipped with the decision-making skills needed to undertake an adaptive management methodology (Gregory et al. 2006). Additionally, economic, social and political priorities can interfere with adaptive management decisions (Gregory et al. 2006). For this reason, adaptive management should be a social process as well as scientific, focusing on institutional strategies while implementing experimental management techniques (Resilience Alliance 2010).

Natural Resource Management

The term natural resource management is frequently used when dealing with a particular resource for human use rather than managing the whole ecosystem (Kellert et al. 2000). A main objective of natural resources management is the sustainability for future generations, which appoints ecosystem managers to balance natural resources exploitation and conservation over long-term timeframe (Ascher 2001). The balanced relationship of each resource in an ecosystem is subject to change at different spatial and temporal scales (Boyce & Haney 1997). Dimensions such as, watersheds, soils, flora and fauna, need to be considered individually and on a landscape level. A variety of natural resources are utilized for food, medicine, energy and shelter (Chapin, Kofinas & Folke 2009).

The ecosystem management concept is based on the relationship between sustainable resource maintenance and human demand for use of natural resources (Boyce & Heney, 1997). Therefore, socioeconomics factors significantly affect natural resource management (Kellert et al. 2000). The goal of a natural resource manager is to fulfill the demand for a given resource without causing harm to the ecosystem, or jeopardizing the future of the resource (Grimble & Wellard 1997). Partnerships between ecosystem managers, natural resource managers and stakeholders should be encouraged in order to promote a more sustainable use of limited natural resources (Cork, Stoneham & Lowe 2007).

Natural resource managers must initially measure the overall integrity of the ecosystem they are involved in. If the ecosystem supporting resources is healthy, managers can decide on the ideal amount of resource extraction, while leaving enough to allow the resource to replenish itself for subsequent harvests (Grimble & Wellard 1997). Historically, some natural resources have experienced limited human disturbance and therefore have been able to subsist naturally . However, some ecosystems such as forests, which typically provide considerable timber resources; have sometimes undergone successful reforestation processes and consequently have accommodated the needs of future generations. A successfully managed resource, will provide for current demand while leaving enough to repopulate and provide for future demand.

Human populations have been increasing rapidly, introducing new stressors to ecosystems, such as climate change and influxes of invasive species. As a result, the demand for natural resources is unpredictable (Chapin, Kofinas & Folke 2009). Although ecosystem changes may occur gradually, the cumulative changes can have negative effects for humans and wildlife (Ascher 2001). Geographic Information Systems (GIS) and Remote Sensing Applications can be used to monitor and evaluate natural resources by mapping them in local and global scales. These tools will continue to be highly beneficial in natural resources management (Boyce & Haney 1997).

Strategic Management

Strategic management encourages the establishment of goals that will benefit the ecosystem while keeping socioeconomic and politically relevant issues in mind (Brussard et al. 1998). Strategic management differs from other types of ecosystem management because it keeps stakeholders involved and relies on their input to develop the best management strategy for an ecosystem. Similarly to other modes of ecosystem management, this method places a high level of importance on evaluating and reviewing any changes, progress or negative impacts and prioritizes flexibility in adapting management protocols as a result of new information (Shmelev and Powell 2006).

Landscape Level Conservation

Landscape level conservation is a method that considers wildlife needs at a broader landscape level scale when implementing conservation initiatives (AWF). This approach to ecosystem management involves the consideration of broad scale interconnected ecological systems that acknowledges the whole scope of an environmental problem (Boyce & Haney 1997). In a human–dominated world, weighing the landcape requirements of wildlife versus the needs of humans is a complicated matter (Opdam & Wascher 2003).

Landscape level conservation is carried out in a number of ways. Wildlife corridors, for example, are connections between otherwise isolated habitat patches that are proposed as a solution to habitat fragmentation (Hudgens & Haddad 2003). In some landscape level conservation approaches, a key species vulnerable to landscape alteration is identified and its habitat requirements are assessed in order to identify the best option for protecting their ecosystem (Lambeck 1997). However, lining up the habitat requirements of numerous species in an ecosystem can be difficult, which is why more comprehensive approaches to further understand these variations have been considered in landscape level conservation (Vos 2010).

Human-induced environmental degradation is an increasing problem globally, which is why landscape level ecology plays an important role in ecosystem management (Velazquez et al. 2003). Traditional conservation methods targeted at individual species need to be modified to include the maintenance of wildlife habitats through consideration of both human-induced and natural environmental factors (Velazquez et al. 2003).

Command and Control Management

Command and control management utilizes a linear problem solving approach where a perceived problem is solved through controlling devices such as laws, threats, contracts and/or agreements (Holling & Meffe 1996). This top-down approach is used across many disciplines and works best with problems that are relatively simple, well-defined and work in terms of cause and effect (Knight & Meffe 1997). The application of command and control management has often attempted to control nature in order to improve product extractions, establish predictability and reduce threats (Holling & Meffe 1996). Some obvious examples of command and control management actions include: the use of herbicides and pesticides to safeguard crops in order to harvest more products; the culling of predators in order to obtain larger, more reliable game species; and the safeguarding of timber supply, by suppressing forest fires .

Attempts at command and control management often backfire (a literal problem in forests that have been 'protected' from fire by humans and are subsequently full of fuel build-up) in ecosystems due to their inherent complexities. Consequently there has been a transition away from command and control management due to many undesirable outcomes and a stronger focus has been placed on more holistic approaches that focus on adaptive management and finding solutions through partnerships (Knight & Meffe 1997).

Ecosystem-based Management

Ecosystem-based management is an environmental management approach that recognizes the full array of interactions within an ecosystem, including humans, rather than considering single issues, species, or ecosystem services in isolation.

Terrestrial ecosystem-based management (often referred to as ecosystem management) came into its own during the conflicts over endangered species protection (particularly the northern spotted owl), land conservation, and water, grazing and timber rights in the western United States in the 1980s and 1990s.

Interest in ecosystem-based management in the marine realm has developed more recently, in response to increasing recognition of the declining state of fisheries and ocean ecosystems.

The systemic origins of ecosystem-based management are rooted in the ecosystem management policy applied to the Great Lakes of North America in the late 1970s. The legislation created, the "Great Lakes Basin and the Great Lakes Water Quality Agreement of 1978", was based on the claim that "no park is an island", with the purpose to show how strict protection of the area is not the best method for preservation. This type of management system was however an idea that began long before and evolved through the testing and challenging of common ecosystem management practices. Before its complete synthesis, the management system's historical development can be traced back to the 1930s. During this time, the scientific communities who studied ecology realized that current approaches to the management of national parks did not provide effective protection of the species within. In 1932, The Ecological Society of America's Committee for the Study of Plant and Animal Communities recognized that US national parks needed to protect all the ecosystems contained within the park in order to create an inclusive and fully functioning sanctuary, and be prepared to handle natural fluctuations in its ecology. Also the committee explained the importance for interagency cooperation and improved public education, as well as challenged the idea that proper park management would "improve" nature (Grumbine 1994). These ideas became the foundation of modern ecosystem-based management.

As the understanding of how to manage ecosystems shifted, new tenets of the management system were produced. Biologists George Wright and Ben Thompson accounted for the size and boundary limitations of parks and contributed to the re-structuring of how park lines were drawn. They explained how large mammals for example could not be supported within the restricted zones of a national park and in order to protect these animals and their ecosystems a new approach would be needed (Wright and Thomson 1935). Other scientists followed suit, but none were successful in establishing a well-defined ecosystem-based management approach. In 1979, the importance of ecosystem-based management resurfaced in ecology from two biologists: John and Frank Craighead. The Craigheads found that grizzly bears of Yellowstone National Park could not sustain a population if only allowed to live within park boundaries. This reinforced the idea that a broader definition of what defines an ecosystem needed to be created, suggesting that it be based on the biotic requirements of the largest mammal present. The idea of ecosystem-based management began to catch on and projects throughout American National Parks reflected the idea of protecting an ecosystem in its entirety and not based on legal or ecological restrictions as previously used. Jim Agee and Darryll Johnson published a book-length report on managing ecosystems in 1988 explaining the theoretical framework management. While they did not fully embrace ecosystem-based management by still calling for "ecologically defined boundaries", they stated the importance of "clearly stated management goals, interagency cooperation, monitoring of management results, and leadership at the national policy levels". Most importantly they demanded the recognition of human influence. It was argued that scientists must keep in mind the "complex social context of their work" and always be moving towards "socially desirable conditions" (Grumbine 1994). This need to understand the social aspects of scientific management is the fundamental step from ecological management to ecosystem-based management. Although it continues to become recognized, a debate over ecosystem-based management continues. Grumbine (1994) believes, while the approach has evolved, it has not been fully incorporated into management practices because the most effective forms of it have yet to be seen. He articulates that the current ecological climate calls for the most holistic approach of ecological management. This is in part due to the rapid decline in biodiversity and because of the constant state of flux in societal and political views of nature. Conflicts over public interest and understanding of the natural world have created social and political climates that require interagency cooperation, which stands as a backbone for ecosystem-based management.

Implementation

Because ecosystem-based management is applied to large, diverse areas encompassing an array of interactions between species, ecosystem components, and humans, it is often perceived as a complex process that is difficult to implement. In addition uncertainty is common and predictions are difficult (Slocombe 1998b). However in light of significant ecosystem degradation, there is a need for a holistic approach that combines environmental knowledge and co-ordination with governing agencies to initiate, sustain and enforce habitat and species protection, and include public education and involvement. As a result, ecosystem-based management will likely be increasingly used in the future as a form of environmental management. Some suggestions for implementing ecosystem-based management and what the process may involve are as follows:

Goals and Objectives

Defining clear and concise goals for ecosystem-based management is one of the most important steps in effective ecosystem-based management implementation. Goals must move beyond science-based or science-defined objectives to include social and cultural importance. Experts also call for the creation of "suites" of goals. A single, end-all goal cannot be the solution, but instead a combination of goals and their relationships with each other should be the focus. As discussed by Slocombe, goals should be broadly applicable, measurable and readily observable, and ideally be collectively supported in order to be achievable. The idea is to provide direction for both thinking and action and should try to minimize managing ecosystems in a static state. Goals should also be flexible enough to incorporate a measure of uncertainty and be able to evolve as conditions and knowledge change.

This may involve focusing on specific threatening processes, such as habitat loss or introduced invasive species, occurring within an ecosystem. Overall the goals should be integrative, to include the structure, organization and processes of the management of an area. Correct ecosystem-based management should be based in goals that are both "substantive", to explain the aims and importance of protecting an area, and "procedural", to explain how substantive goals will be met. As described by Tallis et al., some steps of ecosystem-based management may include:

Scoping

This step involves the acquisition of data and knowledge from various sources in order to provide a thorough understanding of critical ecosystem components. Sources may include literature, informal sources such as aboriginal residents, resource users, and/or environmental experts. Data may also be gained through statistical analyses, simulation models, or conceptual models.

Defining Indicators

Ecological indicators are useful for tracking or monitoring an ecosystem's status and can provide feedback on management progress. Examples may include the population size of a species or the levels of toxin present in a body of water. Social indicators may also be used such as the number or types of jobs within the environmental sector or the livelihood of specific social groups such as indigenous peoples.

Setting Thresholds

Tallis et al. (2010) suggest setting thresholds for each indicator and setting targets that would represent a desired level of health for the ecosystem. Examples may include species composition within an ecosystem or the state of habitat conditions based on local observations or stakeholder interviews. Thresholds can be used to help guide management, particularly for a species by looking at the conservation status criteria established by either state or federal agencies and using models such as the minimum viable population size.

Risk Analysis

A range of threats and disturbances, both natural and human, often can affect indicators. Risk is defined as the sensitivity of an indicator to an ecological disturbance. Several models can be used to assess risk such as population viability analysis.

Monitoring

Evaluating the effectiveness of the implemented management strategies is very important in determining how management actions are affecting the ecosystem indicators. Evaluation: This final step involves monitoring and assessing data to see how well the management strategies chosen are performing relative to the initial objectives stated. The use of simulation models or multi-stakeholder groups can help to assess management.

It is important to note that many of these steps for implementing ecosystem-based management are limited by the governance in place for a region, the data available for assessing ecosystem status and reflecting on the changes occurring, and the time frame in which to operate.

Challenges

Because ecosystems differ greatly and express varying degrees of vulnerability, it is difficult to apply a functional framework that can be universally applied. These outlined steps or components of ecosystem-based management can, for the most part, be applied to multiple situations and are only suggestions for improving or guiding the challenges involved with managing complex issues. Because of the greater amount of influences, impacts, and interactions to account for, problems, obstacles and criticism often arise within ecosystem-based management.

The first commonly defined challenge is the need for meaningful and appropriate management units. These units must be broad and contain value for people in and outside of the protected area. For example Aberley suggests the use of "bioregions" as management units, which can allow peoples involvement with that region to come through. To define management units as inclusive regions rather that exclusive ecological zones would prevent further limitations created by narrow or restricting political and economic policy created from the units. Slocombe suggests that better management units should be flexible and build from existing units and that the biggest challenge is creating truly effect units for managers to compare against.

Another issue is in the creation of administrative bodies. They should operate as the essence of ecosystem-based management, working together towards mutually agreed upon goals. Gaps in administration or research, competing objectives or priorities between management agencies and governments due to overlapping jurisdictions, or obscure goals such as sustainability, ecosystem integrity, or biodiversity can often result in fragmented or weak management (Slocombe 1998b). In addition, limited knowledge of ecosystem components and function and time constraints that can often limit objectives to only those that can be addressed in the short-term (Slocombe 1993; Tallis et al. 2010). The most challenging issue facing ecosystem-based management is that there exists little knowledge about the system and its effectiveness. With limited resources available on how to implement the system it is hard to find support for its use (Slocombe 1998b).

Criticism of ecosystem-based management include its reliance on analogy and comparisons, too broadly applied frameworks, its overlap with or duplication of other methods such as ecosystem management, environmental management, or integrated ecosystem assessment, its vagueness in concepts and application, and its tendency to ignore historical, evolutionary or individual factors that may heavily influence ecosystem functioning.

Ecosystem-based management is seen as a critical planning and management framework for conserving or restoring ecosystems though it is still not widely implemented. An ecosystem approach addresses many relationships across spatial, biological, and organizational scales and is a goal-driven approach to restoring and sustaining ecosystems and functions. In addition, ecosystem-based management involves community influence as well as planning and management from local, regional and national government bodies and management agencies. All must be in collaboration in order to develop a desired future of ecosystem conditions, particularly where ecosystems have undergone radical degradation and change. To move forward, ecosystem-based management should be approached through adaptive management, allowing flexibility and inclusiveness to deal with constant environmental, societal, and political change.

Marine Systems

Ecosystem-based management for marine environments moves away from the traditional strategies in which single species and single sectors are managed individually; rather it is an integrated approach which considers all key activities, particularly anthropogenic, that affect marine environments. The objective is to ensure sustainable ecosystems, thus protecting the resources and services they provide.

In recent years there has been increasing recognition of disruption to marine ecosystems resulting from climate change, overfishing, nutrient and chemical pollution from land runoff, coastal development, bycatch, habitat destruction and other human activities.

There are very clear links between human activities and marine ecosystem functioning; this has become an issue of high importance because there are many services provided by marine ecosystems that are declining as a result of these impacts. These services include the provision of food, fuel, mineral resources, pharmaceuticals, as well as opportunities for recreation, trade, research and education.

Guerry (2005) has identified an urgent need to improve the management of these declining ecosystems, particularly in coastal areas, to ensure a sustainable future. Human communities depend on marine ecosystems for important resources, but without holistic management these ecosystems are likely to collapse.

It has been suggested that the degradation of marine ecosystems is largely the result of poor governance and that new approaches to management are required (Olsson et al. 2008). The Pew Oceans Commission (POC 2003) and the US Commission of Ocean Policy have indicated the importance of moving from current piecemeal management to a more integrated ecosystem-based approach.

Key Elements

Connections: At its core, ecosystem-based management is about acknowledging connections, including the linkages between marine ecosystems and human societies, economies and institutional systems, as well as those among various species within an ecosystem and among ocean places that are linked by the movement of species, materials, and ocean currents.

The more information we can gather about an ecosystem and all of the interconnected factors which affect it, the more capable we will be of better managing that system.

Cumulative Impacts

Ecosystem-based management focuses on how individual actions affect the ecosystem services that flow from coupled social-ecological systems in an integrated fashion, rather than considering these impacts in a piecemeal manner. Loss of biodiversity in marine ecosystems is an example of how cumulative effects from different sectors can impact on an ecosystem in a compounding way. Overfishing, coastal development, filling and dredging, mining and other human activities all contribute to the loss of biodiversity and therefore degradation of the ecosystem.

Interactions Between Sectors

The only way to deal with the cumulative effects of human influences on marine ecosystems is for various contributing sectors to set common goals for the protection or management of ecosystems. While some policies may only affect a single sector, others may affect multiple sectors. A policy for the protection of endangered marine species, for example, could affect recreational and commercial fisheries, mining, shipping and tourism sectors to name a few. More effective ecosystem management would result from the collective adoption of policies by all sectors, rather than each sector creating their own isolated policies.

Changing Public Perceptions

Not all members of the public will be properly informed, or be fully aware, of current threats to marine ecosystems and it is therefore important to change public perceptions by informing people about these issues. It is important to consider the interest of the public when making decisions about ocean management and not just those who have a material interest because community support is needed by management agencies in order to make decisions. The Great Barrier Reef Marine Park Authority (GBRMPA) faced the issue of poor public awareness in their proposed management strategy which included no-take fishing zones. They addressed this problem by starting a 'reef under pressure' information campaign to prove to the public that the Great Barrier Reef is under threat from human disturbances, and in doing so were successful in gaining public support.

Bridging Science and Policy

To ensure that all key players are on the same page, it is important to have communication between managers, resource users, scientists, government bodies and other stakeholders (Leslie and McLeod 2007). Proper engagement between these groups will enable the development of management initiatives that are realistic and enforceable (Leslie and McLeod 2007) as well as effective for ecosystem management.

Embracing Change

Coupled social-ecological systems are constantly changing in ways that cannot be fully predicted or controlled. Understanding the resilience of ecosystems, i.e. the extent to which they can maintain structure, function, and identity in the face of disturbance, can enable better prediction of how ecosystems will respond to both natural and anthropogenic perturbations, and to changes in environmental management.

Multiple Objectives

Ecosystem-based management focuses on the diverse benefits provided by marine systems, rather than on single ecosystem services. Such benefits or services include vibrant commercial and recreational fisheries, biodiversity conservation, renewable energy from wind or waves and coastal protection.

Learning and Adaptation

Because of the lack of control and predictability of coupled social-ecological systems, an adaptive management approach is recommended.

Other Examples

Great Bear Rainforest – Canada: The Land and Resource Management Planning (LRMP) was implemented by the British Columbia Government (Canada) in the mid-1990s in the Great Bear Rainforest in order to establish a multiparty land-use planning system. The aim was to "maintain the ecological integrity of terrestrial, marine and freshwater ecosystems and achieve high levels of human well-being". The steps described in the programme included: protect old-growth forests, maintain forest structure at the stand level, protect threatened and endangered species and ecosystems, protect wetlands and apply adaptive management. MacKinnon (2008) highlighted that the main limitation of this programme was the social and economic aspects related to the lack of orientation to improve human well-being.

The Great Lakes - Canada and United States

A Remedial Action Plan (RAP) was created during the Great Lakes Water Quality Agreement that implemented ecosystem-based management). The transition, according to the authors, from "a narrow to a broader approach " was not easy because it required the cooperation of both the Canadian and American governments. This meant different cultural, political and regulatory perspectives were involved with regards to the lakes. Hartig et al. described eight principles required to make the implementation of ecosystem-based management efficacious: "broad-based stakeholder involvement; commitment of top leaders; agreement on information needs and interpretation; action planning within a strategic framework; human resource development; results and indicators to measure progress; systematic review and feedback; and stakeholder satisfaction".

Sustainable Forest Management

Sustainable forest management (SFM) is the management of forests according to the principles of sustainable development. Sustainable forest management uses very broad social, economic and environmental goals. A range of forestry institutions now practice various forms of sustainable forest management and a broad range of methods and tools are available that have been tested over time.

The "Forest Principles" adopted at The United Nations Conference on Environment and Development (UNCED) in Rio de Janeiro in 1992 captured the general international understanding of sustainable forest management at that time. A number of sets of criteria and indicators have since been developed to evaluate the achievement of SFM at both the country and management unit level. These were all attempts to codify and provide for independent assessment of the degree to which the broader objectives of sustainable forest management are being achieved in practice. In 2007, the United Nations General Assembly adopted the Non-Legally Binding Instrument on All Types of Forests. The instrument was the first of its kind, and reflected the strong international commitment to promote implementation of sustainable forest management through a new approach that brings all stakeholders together.

Definition

A definition of SFM was developed by the Ministerial Conference on the Protection of Forests in Europe (MCPFE), and has since been adopted by the Food and Agriculture Organization (FAO). It defines sustainable forest management as: The stewardship and use of forests and forest lands in a way, and at a rate, that maintains their biodiversity, productivity, regeneration capacity, vitality and their potential to fulfill, now and in the future, relevant ecological, economic and social functions, at local, national, and global levels, and that does not cause damage to other ecosystems. In simpler terms, the concept can be described as the attainment of balance – balance between society's increasing demands for forest products and benefits, and the preservation of forest health and diversity. This balance is critical to the survival of forests, and to the prosperity of forest-dependent communities. For forest managers, sustainably managing a particular forest tract means determining, in a tangible way, how to use it today to ensure similar benefits, health and productivity in the future. Forest managers must assess and integrate a wide array of sometimes conflicting factors commercial and non commercial values, environmental considerations, community needs, even global impact – to produce sound forest plans. In most cases, forest managers develop their forest plans in consultation with citizens, businesses, organizations and other interested parties in and around the forest tract being managed. The tools and visualization have been recently evolving for better management practices.

Because forests and societies are in constant flux, the desired outcome of sustainable forest management is not a fixed one. What constitutes a sustainably managed forest will change over time as values held by the public change.

Criteria and Indicators

Criteria and indicators are tools which can be used to conceptualise, evaluate and implement sustainable forest management. Criteria define and characterize the essential elements, as well as a set of conditions or processes, by which sustainable forest management may be assessed. Periodically measured indicators reveal the direction of change with respect to each criterion.

Criteria and indicators of sustainable forest management are widely used and many countries produce national reports that assess their progress toward sustainable forest management. There are nine international and regional criteria and indicators initiatives, which collectively involve more than 150 countries. Three of the more advanced initiatives are those of the Working Group on Criteria and Indicators for the Conservation and Sustainable Management of Temperate and Boreal Forests (also called the Montreal Process), Forests Europe, and the International Tropical Timber Organization. Countries who are members of the same initiative usually agree to produce reports at the same time and using the same indicators. Within countries, at the management unit level, efforts have also been directed at developing local level criteria and indicators of sustainable forest management. The Center for International Forestry Research, the International Model Forest Network and researchers at the University of British Columbia have developed a number of tools and techniques to help forest-dependent communities develop their own local level criteria and indicators. Criteria and Indicators also form the basis of third-party forest certification programmes such as the Canadian Standards Association's Sustainable Forest Management Standards and the Sustainable Forestry Initiative Standard.

There appears to be growing international consensus on the key elements of sustainable forest management. Seven common thematic areas of sustainable forest management have emerged based on the criteria of the nine ongoing regional and international criteria and indicators initiatives. The seven thematic areas are:

- Extent of forest resources
- Biological diversity
- Forest health and vitality
- Productive functions and forest resources

- Protective functions of forest resources
- Socio-economic functions
- Legal, policy and institutional framework.

This consensus on common thematic areas (or criteria) effectively provides a common, implicit definition of sustainable forest management. The seven thematic areas were acknowledged by the international forest community at the fourth session of the United Nations Forum on Forests and the 16th session of the Committee on Forestry. These thematic areas have since been enshrined in the Non-Legally Binding Instrument on All Types of Forests as a reference framework for sustainable forest management to help achieve the purpose of the instrument. On January 5, 2012, the Montreal Process, Forest Europe, the International Tropical Timber Organization, and the Food and Agriculture Organization of the United Nations, acknowledging the seven thematic areas, endorsed a joint statement of collaboration to improve global forest related data collection and reporting and avoiding the proliferation of monitoring requirements and associated reporting burdens.

Ecosystem Approach

The Ecosystem Approach has been prominent on the agenda of the Convention on Biological Diversity (CBD) since 1995 . The CBD definition of the Ecosystem Approach and a set of principles for its application were developed at an expert meeting in Malawi in 1995, known as the Malawi Principles. The definition, 12 principles and 5 points of "operational guidance" were adopted by the fifth Conference of Parties (COP5) in 2000. The CBD definition is as follows

The ecosystem approach is a strategy for the integrated management of land, water and living resources that promotes conservation and sustainable use in an equitable way. Application of the ecosystem approach will help to reach a balance of the three objectives of the Convention. An ecosystem approach is based on the application of appropriate scientific methodologies focused on levels of biological organization, which encompasses the essential structures, processes, functions and interactions among organisms and their environment. It recognizes that humans, with their cultural diversity, are an integral component of many ecosystems.

Sustainable forest management was recognized by parties to the Convention on Biological Diversity in 2004 (Decision VII/11 of COP7) to be a concrete means of applying the Ecosystem Approach to forest ecosystems. The two concepts, sustainable forest management and the ecosystem approach, aim at promoting conservation and management practices which are environmentally, socially and economically sustainable, and which generate and maintain benefits for both present and future generations. In Europe, the MCPFE and the Council for the Pan-European Biological and Landscape Diversity Strategy (PEBLDS) jointly recognized sustainable forest management to be consistent with the Ecosystem Approach in 2006.

Independent Certification

Growing environmental awareness and consumer demand for more socially responsible businesses helped third-party forest certification emerge in the 1990s as a credible tool for communicating the environmental and social performance of forest operations. There are many potential users of certification, including: forest managers, scientists, policy makers, investors, environmental advocates, business consumers of wood and paper, and individuals. With third-party forest certification, an independent organization develops standards of good forest management, and independent auditors issue certificates to forest operations that comply with those standards. Forest certification verifies that forests are well-managed—as defined by a particular standard—and chain-of-custody certification tracks wood and paper products from the certified forest through processing to the point of sale. This rise of certification led to the emergence of several different systems throughout the world. As a result, there is no single accepted forest management standard worldwide, and each system takes a somewhat different approach in defining standards for sustainable forest management.

In its 2009–2010 Forest Products Annual Market Review United Nations Economic Commission for Europe/Food and Agriculture Organization stated: "Over the years, many of the issues that previously divided the (certification) systems have become much less distinct. The largest certification systems now generally have the same structural programmatic requirements. Third-party forest certification is an important tool for those seeking to ensure that the paper and wood products they purchase and use come from forests that are well-managed and legally harvested. Incorporating third-party certification into forest product procurement practices can be a centerpiece for comprehensive wood and paper policies that include factors such as the protection of sensitive forest values, thoughtful material selection and efficient use of products.

There are more than 50 certification standards worldwide, addressing the diversity of forest types and tenures. Globally, the two largest umbrella certification programmes are:

- Programme for the Endorsement of Forest Certification (PEFC)
- Forest Stewardship Council (FSC)

The area of forest certified worldwide is growing slowly. PEFC is the world's largest forest certification system, with more than two-thirds of the total global certified area certified to its Sustainability Benchmarks. In North America, there are three certification standards endorsed by PEFC – the Sustainable Forestry Initiative, the Canadian Standards Association's Sustainable Forest Management Standard, and the American Tree Farm System. FSC has five standards in North America – one in the United States and four in Canada.

While certification is intended as a tool to enhance forest management practices throughout the world, to date most certified forestry operations are located in Europe and North America. A significant barrier for many forest managers in developing countries is that they lack the capacity to undergo a certification audit and maintain operations to a certification standard.

Forest Governance

Although a majority of forests continue to be owned formally by government, the effectiveness of forest governance is increasingly independent of formal ownership. Since neo-liberal ideology in the 1980s and the emanation of the climate change challenges, evidence that the state is failing to effectively manage environmental resources has emerged. Under neo-liberal regimes in the developing countries, the role of the state has diminished and the market forces have increasingly taken over the dominant socio-economic role. Though the critiques of neo-liberal policies have maintained that market forces are not only inappropriate for sustaining the environment, but are in fact a major cause of environmental destruction. Hardin's tragedy of the common (1968) has shown that the people cannot be left to do as they wish with land or environmental resources. Thus, decentralization of management offers an alternative solution to forest governance.

The shifting of natural resource management responsibilities from central to state and local governments, where this is occurring, is usually a part of broader decentralization process. According to Rondinelli and Cheema (1983), there are four distinct decentralization options: these are: (i) Privatization – the transfer of authority from the central government to non-governmental sectors otherwise known as market-based service provision, (ii) Delegation – centrally nominated local authority, (iii) Devolution – transfer of power to locally acceptable authority and (iv) Deconcentration – the redistribution of authority from the central government to field delegations of the central government. The major key to effective decentralization is increased broad-based participation in local-public decision making. In 2000, the World Bank report reveals that local government knows the needs and desires of their constituents better than the national government, while at the same time, it is easier to hold local leaders accountable. From the study of West African tropical forest, it is argued that the downwardly accountable and/or representative authorities with meaningful discretional powers are the basic institutional element of decentralization that should lead to efficiency, development and equity. This collaborates with the World Bank report in 2000 which says that decentralization should improve resource allocation, efficiency, accountability and equity "by linking the cost and benefit of local services more closely". Many reasons point to the advocacy of decentralization of forest. (i) Integrated rural development projects often fail because they are top-down project that did not take local people's needs and desire into account. (ii) National government sometimes have legal authority over vast forest area that they cannot control, thus, many protected area project result in increased biodiversity loss and greater social conflict. Within the sphere of forest management, as state earlier, the most effective option of decentralization is "devolution"-the transfer of power to locally accountable authority. However, apprehension about local governments is not unfounded. They are often short of resources, may be staffed by people with low education and are sometimes captured by local elites who promote clientelist relation rather than democratic participation. Enters and Anderson (1999) point that the result of community-based projects intended to reverse the problems of past central approaches to conservation and development have also been discouraging.

Broadly speaking, the goal of forest conservation has historically not been met when, in contrast with land use changes; driven by demand for food, fuel and profit. It is necessary to recognized and advocate for better forest governance more strongly given the importance of forest in meeting basic human needs in the future and maintaining ecosystem and biodiversity as well as addressing climate change mitigation and adaptation goal. Such advocacy must be coupled with financial incentives for government of developing countries and greater governance role for local government, civil society, private sector and NGOs on behalf of the "communities".

Sustainable Land Management

Sustainable Land Management is a knowledge-based procedure that aims at integrating the management of land, water, biodiversity, and other environmental resources to meet human needs while sustaining ecosystem services and livelihoods. The term sustainable land management is used, for example, in regional planning and soil or environmental protection but also in property and estate management.

Examples

The World Bank defines Sustainable Land Management as a process in a charged environment between environmental protection and the guarantee claim of ecosystem services on the one hand. On the other hand it is about productivity of agriculture and forestry with respect to demographic growth and increasing pressure in land use. *"SLM is defined as a knowledge-based procedure that helps integrate land, water,* biodiversity, and environmental management (including input and output externalities) to meet rising food and fibre demands while sustaining ecosystem services and livelihoods.

SLM is necessary to meet the requirements of a growing population. Improper land management can lead to land degradation and a significant reduction in the productive and service (biodiversity niches, hydrology, carbon sequestration) functions of watersheds and landscapes."

The United Nations Economic Commission for Europe (UNECE) applies the term in a much wider context. Besides agriculture and forestry they include the mineral extraction sector, property and estate management.

"Land management is the process by which the resources of land are put to good effect. It covers all activities concerned with the management of land as a resource both from an environmental and from an economic perspective. It can include farming, mineral extraction, property and estate management, and the physical planning of towns and the countryside."

In the course of national politics and programmes only few European states use the terminology "sustainable land management". Here Australia and New Zealand are to be mentioned as both countries have agreed on sustainable land management with respect to climate change as part of their government programmes.

In the European context the definition of the European Network for Land Use Management for Sustainable European Cities (LUMASEC) may be used as a reference. It emphasizes the inter- and transdisciplinary cooperation on sustainable land management:

"As management is the human activity meaning the action of people working together in the aim to accomplish desired goals, land use management is a process of managing use and development of land, in which spatial, sector-oriented and temporary aspects of urban policy are coordinated. Resources of land are used for different purposes, which may produce conflicts and competitions, and land use management has to see those purposes in an integrated way. Therefore, land management covers the debate about norms and visions driving the policy-making, sector-based planning both in the strategic and more operative time spans, spatial integration of sectoral issues, decision-making, budgeting, implementation of plans and decisions and the monitoring of results and evaluation of impacts."

Research Examples on Sustainable Land Management

Since 2010, the German Federal Ministry of Education and Research (BMBF) funds an international research programme on "sustainable land management". Module A of the programme investigates interactions between land management, climate change and ecosystem services. It includes projects operating in South America, Africa, Europe, Central and South Asia. Module B seeks "innovative system solutions" in 13 projects with a Central European focus.

Chapter 5

Nature and Human Induced Environmental Hazards

Biodiversity is the extent of variation of life forms within a given species, ecosystem, biome, or an entire planet. Its depletion is termed as biodiversity depletion, i.e., depletion of endemic flora and fauna of a particular region. The loss of biodiversity is a significant issue for scientists and policy-makers and the topic is finding its way into living rooms and classrooms. Species are becoming extinct at the fastest rate known in geological history and most of these extinctions have been tied to human activity. Habitat loss and destruction, usually as a direct result of human activity and population growth, is a major force in the loss of species, populations, and ecosystems.

Alterations in ecosystem composition, such as the loss or decline of a species, can lead to a loss of biodiversity. For example, efforts to eliminate coyotes in the canyons of southern California are linked to decreases in song bird populations in the area. As coyote populations were reduced, the populations of their prey, primarily raccoons, increased. Since raccoons eat bird eggs, fewer coyotes led to more raccoons eating more eggs, resulting in fewer song birds. The introduction of exotic (non native) species can disrupt entire ecosystems and impact populations of native plants or animals. These invaders can adversely affect native species by eating them, infecting them, competing with them, or mating with them. The over exploitation (over-hunting, over fishing, or over-collecting) of a species or population can lead to its demise. Human-generated pollution and contamination can affect all levels of biodiversity.

Global climate change can alter environmental conditions. Species and populations may be lost if they are unable to adapt to new conditions or relocate.

Causes of Recent Declines in Biodiversity

The major causes of biodiversity decline are land use changes, pollution, changes in atmospheric CO_2 concentrations, changes in the nitrogen cycle and acid rain, climate alterations, and the introduction of exotic species, all coincident to human population growth. For rainforests, the primary factor is land conversion. Climate will probably change least in tropical regions, and nitrogen problems are not as important because growth in rainforests is usually limited more by low phosphorus levels than by nitrogen insufficiency. The introduction of exotic species is also less of a problem than in temperate areas because there is so much diversity in tropical forests that newcomers have difficulty becoming established (Sala, *et al.*, 2000).

***Human Population Growth*:** The geometric rise in human population levels during the twentieth century is the fundamental cause of the loss of biodiversity. It exacerbates every other factor having an impact on rainforests (not to mention other ecosystems). It has led to an unceasing search for more arable land for food production and livestock grazing, and for wood for fuel, construction, and energy. Previously undisturbed areas (which may or may not be suitable for the purposes to which they are constrained) are being transformed into agricultural or pasture land, stripped of wood, or mined for resources to support the energy needs of an ever-growing human population. Humans also tend to settle in areas of high biodiversity, which often have relatively rich soils and other attractions for human activities. This leads to great threats to biodiversity, especially since many of these areas have numerous endemic species. Balmford, *et al.*, (2001) have demonstrated that human population size in a given tropical area correlates with the number of endangered species, and that this pattern holds for every taxonomic group. Most of the other effects mentioned below are either consequent to the human population expansion or related to it.

The human population was approximately 600,000 million in 1700, and one billion in 1800. Just now it exceeds six billion, and low estimates are that it may reach 10 billion by the mid-21st century and 12 billion by 2100. The question is whether many ecological aspects of biological systems can be sustained under the pressure of such numbers. Can birds continue to migrate, can larger organisms have space (habitat) to forage, can ecosystems survive in anything like their present form, or are they doomed to impoverishment and degradation?

***Habitat Destruction*:** Habitat destruction is the single most important cause of the loss of rainforest biodiversity and is directly related to human population growth. As rainforest land is converted to ranches, agricultural land (and then, frequently, to degraded woodlands, scrubland, or desert), urban areas (*cf.* Brasilia) and other human usages, habitat is lost for forest organisms. Many species are widely distributed and thus, initially, habitat destruction may only reduce local population numbers. Species which are local, endemic, or which have specialized habitats are much more vulnerable to extinction, since once their particular habitat is degraded or converted for human activity, they will disappear. Most of the habitats being destroyed are those which contain the highest levels of biodiversity, such as lowland tropical wet forests. In this case, habitat loss is caused by clearing, selective logging, and burning.

***Pollution*:** Industrial, agricultural and waste-based pollutants can have catastrophic effects on many species. Those species which are more tolerant of pollution will survive; those requiring pristine environments (water, air, food) will not. Thus, pollution can act as a selective agent. Pollution of water in lakes and rivers has degraded waters so that many freshwater ecosystems are dying. Since almost 12% of animals species live in these ecosystems, and most others depend on them to some degree, this is a very serious matter. In developing countries approximately 90% of wastewater is discharged, untreated, directly into waterways.

***Agriculture*:** The dramatic increase in the number of humans during the twentieth century has instigated a concomitant growth in agriculture, and has led to conversion of wildlands to croplands, massive diversions of water from lakes, rivers and underground aquifers, and, at the same time, has polluted water and land resources with pesticides, fertilizers, and animal wastes. The result has been the destruction, disturbance or disabling of terrestrial ecosystems, and polluted, oxygen-depleted and atrophied water resources. Formerly, agriculture in different regions of the world was relatively independent and local. Now, however, much of it has become part of the global exchange economy and has caused significant changes in social organization.

Earlier agricultural systems were integrated with and co-evolved with technologies, beliefs, myths and traditions as part of an integrated social system. Generally, people planted a variety of crops in different areas, in the hope of obtaining a reasonably stable food supply. These systems could only be maintained at low population levels, and were relatively nondestructive (but not always). More recently, agriculture has in many places lost its local character, and has become incorporated into the global economy. This has led to increased pressure on agricultural land for exchange commodities and export goods. More land is being diverted from local food production to "cash crops" for export and exchange; fewer types of crops are raised, and each crop is raised in much greater quantities than before. Thus, ever more land is converted from forest (and other natural systems) for agriculture for export, rather than using land for subsistence crops. The introduction of monocropping and the use of relatively few plants for food and other uses – at the expense of the wide variety of plants and animals utilized by earlier peoples and indigenous peoples – is responsible for a loss of diversity and genetic variability. The native plants and animals adapted to the local conditions are now being replaced with "foreign" (or "exotic") species which require special inputs of food and nutrients, large quantities of water. Such exotic species frequently drive out native species. There is pressure to conform to crop selection and agricultural techniques – all is driven by global markets and technologies.

***Global Warming*:** There is recent evidence that climate changes are having effects on tropical forest ecology. Warming in general (as distinct from the effects of increasing concentrations of CO_2 and other greenhouse gases) can increase primary productivity, yielding new plant biomass, increased organic litter, and increased food supplies for animals and soil flora (decomposers). Temperature changes can also alter the water cycle and the availability of nitrogen and other nutrients. Basically, the temperature variations which are now occurring affect all parts of forest ecosystems, some more than others. These interactions are unimaginably complex. While warming may at first increase net primary productivity (NPP), in the longer run, because plant biomass is increasing, more nitrogen is taken up from the soil and sequestered in the plant bodies. This leaves less nitrogen for the growth of additional plants, so the increase in NPP over time (due to a rise in temperature or CO_2 levels) will be limited by nitrogen availability. The same is probably true of other mineral nutrients. The consequences of warming-induced shifts in the distribution of nutrients will not be seen rapidly, but perhaps only over many years. These events may effect changes in species distribution and other ecosystem processes in complex ways. We know little about the reactions of tropical forests, but they may differ from those of temperate forests.

In tropical forests, warming may be more important because of its effects on evapotranspiration and soil moisture levels than because of nutrient redistribution or NPP (which is already very high because tropical temperatures are close to the optimum range for photosynthesis and there is so much available light energy). And warming will obviously act in concert with other global or local changes – increases in atmospheric CO_2 (which may modify plant chemistry and the water balance of the forest) and land clearing (which changes rainfall and local temperatures), for examples.

Root, *et al.*(2003) have determined that more than 80%of plant and animal species on which they gathered data had undergone temperature-related shifts in physiology. Highland forests in Costa Rica have suffered losses of amphibian and reptile populations which appear to be due to increased warming of montane forests. The golden toad *Bufo periglenes* of Costa Rica has become extinct, at least partly because of the decrease in mist frequency in its cloud forest habitat. The changes in mists appear to be a consequence of warming trends. Other suspected causes are alterations in juvenile growth or maturation rates or sex ratios due to temperature shifts. Parmesan and Yohe (2003), in a statistical analysis, determined that climate change had biological effects on the 279 species which they examined.

The migratory patterns of some birds which live in both tropical and temperate regions during the year seem to be shifting, which is dangerous for these species, as they may arrive at their breeding or wintering grounds at an inappropriate time. Or they may lose their essential interactions with plants which they pollinate or their insect or plant food supplies. Perhaps for these reasons, many migratory species are in decline, and their inability to coördinate migratory clues with climatic actualities may be partly to blame. The great tit, which still breeds at the same time as previously, now misses much of its food supply because its plant food develops at an earlier time of year, before the birds have arrived from their wintering grounds. Also, as temperatures rise, some bird populations have shifted, with lowland and foothill species moving into higher areas. The consequences for highland bird populations are not yet clear. And many other organisms, both plant and animal, are being affected by warming.

An increase in infectious diseases is another consequence of climate change, since the causative agents are affected by humidity, temperature change, and rainfall. Many species of frogs and lizards have declined or disappeared, perhaps because of the increase in parasites occasioned by higher temperatures. As warming continues, accelerating plant growth, pathogens may spread more quickly because of the increased availability of vegetation (a "density" effect) and because of increased humidity under heavier plant cover. As mentioned above, the fungus *Phytophtora cinnamoni* has demolished many *Eucalyptus* forests in Australia. In addition, the geographical range of pathogens can expand when the climate moderates, allowing pathogens to find new, nonresistant hosts. On the other hand, a number of instances of amphibian decline seem to be due to infections with chrytid fungi, which flourish at cooler temperatures. An excellent review of this complex issue may be found in Harvell, *et al.*, (2002).

There may be a link between augmented carbon dioxide levels and marked increase in the density of lianas in Amazonian forests. This relationship is suggested by the fact that growth rates of lianas are highly sensitive to CO_2 levels. As lianas become more dense, tree mortality rises, but mortality is not equal among species because lianas preferentially grow on certain species. Because of this biodiversity may be reduced by increased mortality in some species but not others (Phillips, *et al.*, 2002).

***Forest Fragmentation*:** The fragmentation of forests is a general consequence of the haphazard logging and agricultural land conversion which is occurring everywhere, but especially in tropical forests. When forests are cut into smaller and smaller pieces, there are many consequences, some of which may be unanticipated.

Fragmentation decreases habitat simply through loss of land area, reducing the probability of maintaining effective reproductive units of plant and animal populations. Most tropical trees are pollinated by animals, and therefore the maintenance of adequate pollinator population levels is essential for forest health. When a forest becomes fragmented, trees of many species are isolated because their pollinators cannot cross the unforested areas. Under these conditions, the trees in the fragments will then become inbred and lose genetic variability and vigor. Other species, which have more wide-ranging pollinators, may suffer less from fragmentation. For instance, the pollen of several species of strangler figs (the fruit of which is an essential element in the diets of many animals) is dispersed by wasps over distances as great as 14.2 km (Nason, Herre, & Hamrick, 1998). Thus "breeding units" of these figs are extremely large, comprising hundreds of plants located in huge areas of forest. Isolated fig populations seem to survive and help to maintain frugivore numbers (if not diversity), so long as the number of trees within the range of the wasps does not fall below a critical minimum.

Most species are not so tolerant, however. Animals, particularly large ones, cannot maintain themselves in small fragmented forests. Many large mammals have huge ranges and require extensive areas of intact forest to obtain sufficient food, or to find suitable nesting sites. Additionally, their migrations may be interrupted by fragmentation. These animals are also much more susceptible to hunting in forest fragments, which accounts for much of the decline in animal populations in rainforests. Species extinctions occur more rapidly in fragments, for these reasons, and also because species depend upon each other.

The dissection of forests into fragments in certain parts of the Amazon has led to extreme hunting pressures on peccaries, for instance, and in some places where they are locally extinct, three species of frogs have also disappeared, since they depended upon peccary wallows for breeding ponds. The absence of large predator species leads to imbalances in prey populations, and, since many of the prey species are seed-eaters, to declines in the population levels of many plant species. The prey, now at high population levels, consume most available seeds, leaving few to germinate. On small islands created after dam construction on the Chagres River in Panama, even large seed predators could not survive, and after 70 years, the former mixed tropical forest has become a forest of large-seeded plants only (Terborgh, 1992b). As Terborgh states, and we should attend to this lesson, "Distortions in any link of the interaction chain will induce changes in the remaining links." (p. 289) When forests are cut down or burned, the resulting gaps are too large to be filled in by the normal regeneration processes. This permits the ascendancy of rapid-growing, light-tolerant species and grasses. Large gaps may then be converted to scrub or grassland. The "edge" effect: The cutting of forest into fragments creates many "edges" where previously there was deep forest. Many effects are consequent upon this. Edges are lighter, warmer and windier than the forest interior. These changes in microclimate alter plant reproduction, animal distribution, the biological structure and many other features of the forest.

Tree mortality is much greater near edges, and climax species will be replaced by pioneer species. These effects can be seen as far as one kilometer into the forest. The drier and warmer conditions also make the fragment more flammable, with a concomitant increase in the frequency of fires. Without further stress, the forest may regenerate. However, if the fragment is surrounded by a human-dominated landscape, it may be inhibited from regeneration. This has occurred in certain areas of Brazil, where forest fragments are surrounded by sugar cane and *Eucalyptus* plantations. Thus, species requiring large areas of undisturbed primary forest are sacrificed to the benefit of those species which can exist on forest margins.

Fire is particularly frequent in fragments. Recently, many forests have been subjected to deliberately-set and accidental fires, to which they have little resistance, and to which they are rarely naturally subjected. People often set fire to cut-over areas adjacent to forests to clear them of debris.

These fires often get out of control and burn large areas, extend into the forest interior, and inhibit edge regeneration by killing pioneer forest vegetation. More than 90% of forest fires in certain eastern Amazon forest areas were associated with the edges of forest fragments (Wuethrich, 2000). If conditions remain severe, the forest will recede and be replaced by scrub.

The use of herbicides and the introduction of exotic species into areas surrounding forest fragments are detrimental to forest health. Herbicides blow from cleared agricultural areas into forests, and exotic species introduced by farmers and ranchers spread, often displacing native species. These exotic organisms interrupt the forest ecosystem and, since they have few or no natural enemies in their new environment, they are difficult to eradicate. According to Vitousek (1997), there are many islands where fewer than half of the species are native, and in many other terrestrial environments, more than 20% of species are foreign. These invasions drive the loss of indigenous species.

For unknown reasons, fragmentation leads to the death of large canopy trees, even in the interior of fragments. Canopy trees dominate the forest structure, and they provide fruits and shelter for many animals. The mortality of trees in fragmented patches in Brazil has been found to be twice that of similar trees in the forest interior (Laurance, *et al.*, 2000). Not only that, but tree mortality is confined disproportionately to large trees (an almost 40% increase in mortality). Large trees may be more vulnerable in fragmented forests because they are not as well buffered from wind and natural forces, because there are more tree parasites (lianas), and because they are more subject to dessication at forest edges. Loss of these largest trees has several corollary effects – the alteration of biogeochemical cycles (transpiration, carbon cycles), the reduction of species complexity, and the reduction of fecundity. As mentioned above, large trees are essential habitats and food sources for many other organisms, both plant and animal; they are the source of much of the primary productivity of the forest; and they are responsible for many effects on the water and nutrient cycles. They are irreplaceable in the forest ecosystem.

The fragmentation of forests by logging and agricultural conversion also exaggerates the probability of major epidemics. Pathogens introduced through human activities by land use practices in areas surrounding the forest can be lethal to forest plants and animals.

Rainforests are losing species, not only because of the disappearance of their habitat, but also because essential ecological processes are being interrupted by fragmentation. Fragments are much more easily accessible to human incursions than are intact forests. This leads to a variety of extractive activities within the forest interior. Intensive hunting, by depleting animal populations, inhibits plant reproduction, since many seeds can neither be dispersed, nor flowers be pollinated without them. Where these seed dispersers have been eliminated, are at low population densities, or cannot move between forest fragments, seed dispersal will be very limited, and as a result tree species dependent upon animal dispersers may become locally extinct. In the remnants of the Atlantic forest of Brazil, the seeds of 71% of tree species are dispersed by vertebrates (birds and mammals), and about 48% of these dispersers are birds which are deep-forest dwellers. As this forest becomes more and more fragmented, these birds are disappearing, so eventually the trees dependent upon them will be unable to replace themselves. In some fragments, all large vertebrates (including seed-eaters) have been hunted to extinction, and in some places the fragments are so distant from each other that these animals cannot pass from one to another. The Alagoas curassow, a large fruit-eating bird of this area, is now extinct; many others (toucans, aracaris, guans) are endangered by hunting pressures. Other species are sensitive to disturbance of their environment and they become locally extinct. The tiny bits of Atlantic forest remaining are becoming dominated by trees which are wind or water-pollinated or whose seeds are dispersed by animal species which can tolerate disturbed habitats or edges ("edge species") (Cardoso da Silva and Tabarelli, 2000).

In addition, in fragmented forests, seeds will frequently land in deforested areas (where they are in the open, and exposed to heat, light and desiccation) in which they cannot germinate, and the seedlings cannot survive. In Brazil, three to seven times as many *Heliconia acuminata* seedlings planted in continuous areas of forest germinated as compared to those planted in fragmented areas (Bruna, 1999).

Whatever the explanation for the lower rate of seedling germination in fragmented forests, whether due to inbreeding or other causes, fewer and fewer individuals in fragments grow to adulthood. Those which do will breed, but since populations are small, inbreeding occurs and the downward spiral continues until the population becomes locally extinct. This effect is seen frequently in forest fragments.

***Hunting, Fishing, and Gathering*:** Many forests which appear intact are in fact "empty forests," since most large animals have been hunted to unsustainable levels. These animals are mainly hunted for meat, but also for skins (jaguar, ocelot) or medicinal/chemical properties (poison-arrow frogs, collected to provide poisons for arrow tips, and the midwife toad, which in the Amazon is thought to have medicinal value). Turtles are heavily harvested for meat and their eggs are collected for food almost everywhere in the tropics and subtropics. Asian tropical freshwater turtles are in serious decline because they are extensively hunted for food or for use in traditional Chinese medicines. Thousands of tons of live turtles are caught or sent to China annually, a completely unsustainable level of collection. There are apparently no turtles left in the wild in Vietnam for this reason (Gibbons, *et al.*, 2000). More than 80 species of Asian turtles are at such low population numbers that they will become extinct unless emergency measures – restrictions on international trade, increased habitat protection, captive breeding programmes – are taken immediately.

Some of the hunting is done for subsistence purposes by villagers; some by farmers, miners and loggers, who live in the forest and use forest animals as a major food source; some by commercial hunters to supply urban markets. This is a major source of income in many rural tropical areas. In Gabon alone, a tiny country, 3,600 tons of bushmeat are consumed annually (Tuxill, 1998). The popularity of "bushmeat" in cities and towns located in or close to rainforests is rising. Surveys of bushmeat consumption in Bolivia and Honduras showed that people will eat more bushmeat as their income rises, but when they become more affluent, consumption declines. Consumption declines as well when bushmeat prices rise and the price of alternative sources of protein declines (Wilkie and Godoy, 2000). Part of the remedy for overexploitation of wildlife resources, then, lies with improving the income levels of local residents (so bushmeat becomes less attractive as a protein source), in increasing the costs of hunting, and in lowering the prices of alternative protein sources.

Many animals are trapped for the pet trade (tropical fish, birds, reptiles, monkeys) or for zoos or medical research. A 25-week survey of the Bangkok weekend market found specimens of 225 species of birds (most "protected" by government decree) for sale (Sponsel, Bailey and Headland, 1996). Other animals are trapped for their hides or furs, and some are killed because they live too close to human habitation and impinge on human activities. For instance, ocelots and other small carnivores may be shot when they attempt to prey on chickens or other domestic animals.

Many tropical animals are hunted mercilessly for their value in traditional Asian medicines. Tigers, bears, deer, snakes, and many other animals are near extinction in many places because of this trade. Tigers in India are almost gone and only 3-5000 tigers still exist in the wild anywhere. Many of these animals, or their parts, are smuggled illegally from Southeast Asian (and other) countries to China and other countries with large Chinese populations for these uses.

The effects of hunting are not just on the animals "taken." Many animals which are human prey eat fruits and seeds, and are major seed dispersers in tropical forests, and the seeds of certain species of trees must pass through the gut of an animal in order to germinate. In these ways many tropical plants and trees depend upon animals, for, without them, they will not be able to reproduce. For instance, the seeds of *Inga ingoides*, a South American tree, are dispersed widely by the spider monkey *Ateles paniscus*. Where this monkey is locally extinct (due to hunting pressures), the trees do not "outbreed"; the seeds fall to the forest floor and patches of seedlings of low genetic diversity surround the parent trees (Moore, 2001). This can be very detrimental to forests, which generally have high genetic diversity, because more homogeneous plants are generally less fit. The loss of elephants in African countries due to hunting has led to a loss of reproductive ability in many valuable tree species (Tuxill, 1998).

Fish and aquatic animals are killed indiscriminately by fishing techniques which employ insecticides and/or dynamite. These techniques not only catch the few desired specimens, but kill all of the other animals in the area. Commercial fishing operations are not sensitive to issues of sustainability. They catch as many marketable fish as possible, and intensify their efforts when fish populations drop (declines due in the first place to overfishing). Such unsustainable fishing operations have led and are leading to severe declines in fish in major river systems within tropical rainforests.

Fragmentation may be more serious than previously imagined because the consequences of fragmentation are not static, but progressive. The edges of cut areas do not remain "in place" but gradually recede, further reducing the size of the fragments. Eventually fragments may disappear altogether or undergo ecological collapse.

Why do people heedlessly decimate the precious biodiversity of their planet? Some of them feel they have no economic alternative, while others are driven by the desire for short-term profit. Still others are uncomprehending. Unfortunately, so much of the depredation which is being inflicted upon areas of great biodiversity is, in the long run, and often in the short run, in vain. While tropical forests now occupy less than half of their former range, and much of what remains is damaged or fragmented, the net profit to humanity is slight. Clearing of tropical forests has provided only a relatively small percentage of total agricultural land, since much of the land converted for farms becomes rapidly degraded and is abandoned. Logging results in a one-time profit, mainly to large companies. Ranching is an activity which, on former rainforest land, is uneconomical, requires subsidizing, and is eventually abandoned. But the damage is permanent and the forest irreplaceable, so forest destruction has dire consequences. It degrades aquatic fisheries, causes floods and has many other consequences – so much harm for so little benefit.

Chapter 6

Environmental Pollution: Problems and Control

Environment is the biotic and abiotic surrounding of an organism, or population, and includes particularly the factors that have an influence in their survival, development and evolution. The naked term environment can make reference to different concepts, but it is often used as a short form for the *biophysical environment*. This practice is common, for instance, among governments, that usually name their departments and agencies dealing with the biophysical environment with denominations like Environment Agency. Whereas the expression "*the environment*" is often used to refer to the global environment, usually as referred to humanity, the number of biophysical environments is countless, given that it is always possible to consider an additional living organism that has its own environment.

Life has to be adapted to the conditions of its environment. Temperature, light, humidity, soil nutrients, etc., all influence any species, within any environment. However, life by its activity in turn modifies, in various forms, the conditions themselves. Some long term modifications along the history of our planet have been significant, such as the incorporation of oxygen to the atmosphere. This process consisted in the breakdown of carbon dioxide by anaerobic microorganisms that used the carbon in their metabolism and released the oxygen to the atmosphere. This led to the existence of oxygen-based plant and animal life (Great oxygenation event). Other interactions are more immediate and simple, such as the smoothing effect that forests have on the temperature cycle, compared to neighbouring unforested areas.

Pollution

Pollution is the introduction of contaminants into the natural environment that cause adverse change. Pollution can take the form of chemical substances or energy, such as noise, heat or light. Pollutants, the components of pollution, can be either foreign substances/energies or naturally occurring contaminants. Pollution is often classed as point source or nonpoint source pollution.

Ancient Cultures

Air pollution has always accompanied civilizations. Pollution started from the prehistoric times when man created the first fires. According to a 1983 article in the journal *Science,* "soot found on ceilings of prehistoric caves provides ample evidence of the high levels of pollution that was associated with inadequate ventilation of open fires." The forging of metals appears to be a key turning point in the creation of significant air pollution levels outside the home. Core samples of glaciers in Greenland indicate increases in pollution associated with Greek, Roman and Chinese metal production, but at that time the pollution was comparatively less and could be handled by nature.

Official Acknowledgement

King Edward I of England banned the burning of sea-coal by proclamation in London in 1272, after its smoke became a problem. But the fuel was so common in England that this earliest of names for it was acquired because it could be carted away from some shores by the wheelbarrow. Air pollution would continue to be a problem in England, especially later during the industrial revolution, and extending into the recent past with the Great Smog of 1952. London also recorded one of the earlier extreme cases of water quality problems with the Great Stink on the Thames of 1858, which led to construction of the London sewerage system soon afterward. It was the industrial revolution that gave birth to environmental pollution as we know it today. The emergence of great factories and consumption of immense quantities of coal and other fossil fuels gave rise to unprecedented air pollution and the large volume of industrial chemical discharges added to the growing load of untreated human waste. Chicago and Cincinnati were the first two American cities to enact laws ensuring cleaner air in 1881. Other cities followed around the country until early in the 20th century, when the short lived Office of Air Pollution was created under the Department of the Interior. Extreme smog events were experienced by the cities of Los Angeles and Donora, Pennsylvania in the late 1940s, serving as another public reminder.

Modern Awareness

Pollution became a popular issue after World War II, due to radioactive fallout from atomic warfare and testing. Then a non-nuclear event, The Great Smog of 1952 in London, killed at least 4000 people. This prompted some of the first major modern environmental legislation, The Clean Air Act of 1956. Pollution began to draw major public attention in the United States between the mid-1950s and early 1970s, when Congress passed the Noise Control Act, the Clean Air Act, the Clean Water Act and the National Environmental Policy Act.

Severe incidents of pollution helped increase consciousness. PCB dumping in the Hudson River resulted in a ban by the EPA on consumption of its fish in 1974. Long-term dioxin contamination at Love Canal starting in 1947 became a national news story in 1978 and led to the Superfund legislation of 1980. Legal proceedings in the 1990s helped bring to light hexavalent chromium releases in California—the champions of whose victims became famous. The pollution of industrial land gave rise to the name brownfield, a term now common in city planning. The development of nuclear science introduced radioactive contamination, which can remain lethally radioactive for hundreds of thousands of years. Lake Karachay, named by the Worldwatch Institute as the "most polluted spot" on earth, served as a disposal site for the Soviet Union throughout the 1950s and 1960s. Second place may go to the area of Chelyabinsk U.S.S.R. as the "Most polluted place on the planet".

Nuclear weapons continued to be tested in the Cold War, sometimes near inhabited areas, especially in the earlier stages of their development. The toll on the worst-affected populations and the growth since then in understanding about the critical threat to human health posed by radioactivity has also been a prohibitive complication associated with nuclear power. Though extreme care is practiced in that industry, the potential for disaster suggested by incidents such as those at Three Mile Island and Chernobyl pose a lingering specter of public mistrust. One legacy of nuclear testing before most forms were banned has been significantly raised levels of background radiation. International catastrophes such as the wreck of the Amoco Cadiz oil tanker off the coast of Brittany in 1978 and the Bhopal disaster in 1984 have demonstrated the universality of such events and the scale on which efforts to address them needed to engage. The borderless nature of atmosphere and oceans inevitably resulted in the implication of pollution on a planetary level with the issue of global warming. Most recently the term persistent organic pollutant (POP) has come to describe a group of chemicals such as PBDEs and PFCs among others. Though their effects remain somewhat less well understood owing to a lack of experimental data, they have been detected in various ecological habitats far removed from industrial activity such as the Arctic, demonstrating diffusion and bioaccumulation after only a relatively brief period of widespread use.

A much more recently discovered problem is the Great Pacific Garbage Patch, a huge concentration of plastics, chemical sludge and other debris which has been collected into a large area of the Pacific Ocean by the North Pacific Gyre. This is a less well known pollution problem than the others described above, but none the less has multiple and serious consequences such as increasing wildlife mortality, the spread of invasive species and human ingestion of toxic chemicals. Organizations such as 5 Gyres have researched the pollution and, along with artists like Marina DeBris, are working toward publicizing the issue. Growing evidence of local and global pollution and an increasingly informed public over time have given rise to environmentalism and the environmental movement, which generally seek to limit human impact on the environment.

Forms of Pollution

The major forms of pollution are listed below along with the particular contaminant relevant to each of them:

- Air pollution:- the release of chemicals and particulates into the atmosphere. Common gaseous pollutants include carbon monoxide, sulphur dioxide, chlorofluorocarbons (CFCs) and nitrogen oxides produced by industry and motor vehicles. Photochemical ozone and smog are created as nitrogen oxides and hydrocarbons react to sunlight. Particulate matter, or fine dust is characterized by their micrometre size PM_{10} to $PM_{2.5}$.
- Light pollution:- includes light trespass, over-illumination and astronomical interference.
- Littering:- the criminal throwing of inappropriate man-made objects, unremoved, onto public and private properties.
- Noise pollution:- which encompasses roadway noise, aircraft noise, industrial noise as well as high-intensity sonar.

- Soil contamination occurs when chemicals are released by spill or underground leakage. Among the most significant soil contaminants are hydrocarbons, heavy metals, MTBE, herbicides, pesticides and chlorinated hydrocarbons.
- Radioactive contamination, resulting from 20th century activities in atomic physics, such as nuclear power generation and nuclear weapons research, manufacture and deployment.
- Thermal pollution, is a temperature change in natural water bodies caused by human influence, such as use of water as coolant in a power plant.
- Visual pollution, which can refer to the presence of overhead power lines, motorway billboards, scarred landforms (as from strip mining), open storage of trash, municipal solid waste or space debris.
- Water pollution, by the discharge of wastewater from commercial and industrial waste (intentionally or through spills) into surface waters; discharges of untreated domestic sewage, and chemical contaminants, such as chlorine, from treated sewage; release of waste and contaminants into surface runoff flowing to surface waters (including urban runoff and agricultural runoff, which may contain chemical fertilizers and pesticides); waste disposal and leaching into groundwater; eutrophication and littering.

Pollutants

A pollutant is a waste material that pollutes air, water or soil. Three factors determine the severity of a pollutant: its chemical nature, the concentration and the persistence. A pollutant is a substance or energy introduced into the environment that has undesired effects, or adversely affects the usefulness of a resource. A pollutant may cause long- or short term damage by changing the growth rate of plant or animal species, or by interfering with human amenities, comfort, health, or property values. Some pollutants are biodegradable and therefore will not persist in the environment in the long term. However the degradation products of some pollutants are themselves polluting such as the products DDE and DDD produced from degradation of DDT.

Different Types of Pollutants by Absorptive Capacity

Stock pollutants: Pollutants that the environment has little or no absorptive capacity are called stock pollutants (e.g. persistent synthetic chemicals, non-biodegradable plastics, and heavy metals). Stock pollutants accumulate in the environment over time. The damage they cause increases as more pollutant is emitted, and persists as the pollutant accumulates. Stock pollutants can create a burden for future generations by passing on damage that persists well after the benefits received from incurring that damage have been forgotten.

Fund Pollutants

Fund pollutants are those for which the environment has some absorptive capacity. Fund pollutants do not cause damage to the environment unless the emission rate exceeds the receiving environment's absorptive capacity (e.g. carbon dioxide, which is absorbed by plants and oceans). Fund pollutants are not destroyed, but rather converted into less harmful substances, or diluted/dispersed to non-harmful concentrations.

Notable Pollutants

Notable pollutants include the following groups:

Heavy Metals

A heavy metal is a member of a loosely defined subset of elements that exhibit metallic properties. It mainly includes the transition metals, some metalloids, lanthanides, and actinides. Many different definitions have been proposed—some based on density, some on atomic number or atomic weight, and some on chemical properties or toxicity. The term *heavy metal* has been called a "misinterpretation" in an IUPAC technical report due to the contradictory definitions and its lack of a "coherent scientific basis". There is an alternative term *toxic metal*, for which no consensus of exact definition exists either. As discussed below, depending on context, heavy metal can include elements lighter than carbon and can exclude some of the heaviest metals. Heavy metals occur naturally in the ecosystem with large variations in concentration. In modern times, anthropogenic sources of heavy metals, i.e. pollution, have been introduced to the ecosystem.

Relationship to Living Organisms

Living organisms require varying amounts of "heavy metals". Iron, cobalt, copper, manganese, molybdenum, and zinc are required by humans. Excessive levels can be damaging to the organism. Other heavy metals such as mercury, plutonium, and lead are toxic metals and their accumulation over time in the bodies of animals can cause serious illness. Certain elements that are normally toxic are, for certain organisms or under certain conditions, beneficial. Examples include vanadium, tungsten, and even cadmium.

Heavy metal toxicity can result in damaged or reduced mental and central nervous function, lower energy levels, and damage to blood composition, lungs, kidneys, liver, and other vital organs. Long-term exposure may result in slowly progressing physical, muscular, and neurological degenerative processes that mimic Alzheimer's disease, Parkinson's disease, muscular dystrophy, and multiple sclerosis. Allergies are not uncommon, and repeated long-term contact with some metals (or their compounds) may cause cancer.

Heavy metals have effect on stability of colloids. Prokaryotic and eukaryotic cells are colloids, too. Colloids are sensitive to change of ion or heavy metal concentration and it leads to violation of colloid stability and subsequent disintegration.

Heavy Metal Pollution

Motivations for controlling heavy metal concentrations in gas streams are diverse. Some of them are dangerous to health or to the environment (e.g. mercury, cadmium, lead, chromium), some may cause corrosion (e.g., lead), some are harmful in other ways (e.g. arsenic may pollute catalysts). Within the European community the eleven elements of highest concern are arsenic, cadmium, cobalt, chromium, copper, mercury, manganese, nickel, lead, tin, and thallium, the emissions of which are regulated in waste incinerators. Some of these elements are actually necessary for humans in minute amounts (cobalt, copper, chromium, manganese, nickel) while others are carcinogenic or toxic, affecting, among others, the central nervous system (manganese, mercury, lead, arsenic), the kidneys or liver (mercury, lead, cadmium, copper) or skin, bones, or teeth (nickel, cadmium, copper, chromium).

Heavy metal pollution can arise from many sources but most commonly arises from the purification of metals, e.g., the smelting of copper and the preparation of nuclear fuels. Electroplating is the primary source of chromium and cadmium. Cadmium, lead and zinc are released in tiny particulates as dust from rubber tires on road surfaces; the small size allows these toxic metals to rise on the wind to be inhaled, or transported onto topsoil or edible plants. Through precipitation of their compounds or by ion exchange into soils and muds, heavy metal pollutants can localize and lay dormant, which can have severe effects on the environment. Unlike organic pollutants, heavy metals do not decay and thus pose a different kind of challenge for remediation.

Plants, mushrooms, or microrganisms are occasionally successfully used to remove some heavy metals such as mercury. Plants which exhibit hyper accumulation can be used to remove heavy metals from soils by concentrating them in their bio matter. Some treatment of mining tailings has occurred where the vegetation is then incinerated to recover the heavy metals.

One of the largest problems associated with the persistence of heavy metals is the potential for bioaccumulation and biomagnification causing heavier exposure for some organisms than is present in the environment alone. Coastal fish (such as the smooth toadfish) and seabirds (such as the Atlantic Puffin) are often monitored for the presence of such contaminants.

Medicine

In medical usage, heavy metals are loosely defined and include all toxic metals irrespective of their atomic weight: "heavy metal poisoning" can possibly include excessive amounts of iron, manganese, aluminium, mercury, cadmium, beryllium or such a semimetal as arsenic. This definition excludes bismuth, the heaviest of approximately stable elements, because of its low toxicity. Minamata disease results from mercury poisoning, and itai-itai disease from cadmium poisoning.

Hazardous Materials

Heavy metals in a hazardous materials (or "hazmat") setting are for the most part classified in "Misc." on the UN model hazard class, but they are sometimes labelled as a poison when being transported..

Nuclear Technology

Burnup of nuclear fuel is expressed in gigawatt-days per metric ton of heavy metal, where heavy metal means actinides like thorium, uranium, plutonium, etc., including both fissile and fertile material. It does not include elements such as oxygen that may be bonded to the fuel metals, or cladding materials such as zirconium, which might be considered heavy metals by some other standards.

Persistent Organic Pollutants POP

Persistent organic pollutants (POPs) are organic compounds that are resistant to environmental degradation through chemical, biological, and photolytic processes. Because of this, they have been observed to persist in the environment, to be capable of long-range transport, bioaccumulate in human and animal tissue, biomagnify in food chains, and to have potential significant impacts on human health and the environment. Many POPs are currently or were in the past used as pesticides. Others are used in industrial processes and in the production of a range of goods such as solvents, polyvinyl chloride, and pharmaceuticals. There are a few natural sources of POPs, but most POPs are created by humans in industrial processes, either intentionally or as byproducts.

Compounds

In May 1995, the United Nations Environment Programme Governing Council (GC) decided to begin investigating POPs, initially beginning with a short list of the following twelve POPs, known as the 'dirty dozen': aldrin, chlordane, DDT, dieldrin, endrin, heptachlor, hexachlorobenzene, mirex, polychlorinated biphenyls, polychlorinated dibenzo-p-dioxins, polychlorinated dibenzofurans, and toxaphene. Since then, this list has generally been accepted to include such substances as carcinogenic polycyclic aromatic hydrocarbons (PAHs) and certain brominated flame-retardants, as well as some organometallic compounds such as tributyltin (TBT).

The groups of compounds that make up POPs are also classed as PBTs (Persistent, Bioaccumulative and Toxic) or TOMPs (Toxic Organic Micro Pollutants.)

Chemical Properties

Some of the chemical characteristics of POPs include low water solubility, high lipid solubility, semi-volatility, and high molecular masses. POPs with molecular masses lower than 236 g/mol are less toxic, less persistent in the environment, and have more reversible effects than those with higher molecular masses. POPs are frequently halogenated, usually with chlorine. The more chlorine groups a POP has, the more resistant it is to being broken down over time. One important factor of their chemical properties such as lipid solubility results in the ability to pass through biological phospholipid membranes and bioaccumulate in the fatty tissues of living organisms.

Long-range Transport

POPs released to the environment have been shown to travel vast distances from their original source. Due to their chemical properties, many POPs are semi-volatile and insoluble. These compounds are therefore unable to transport directly through the environment. The indirect routes include attachment to particulate matter, and through the food web. The chemicals' semi-volatility allows them to travel long distances through the atmosphere before being deposited. Thus POPs can be found all over the world, including in areas where they have never been used and remote regions such as the middle of oceans and Antarctica. The chemicals' semi-volatility also means that they tend to volatilize in hot regions and accumulate in cold regions, where they tend to condense and stay. PCBs have been found in precipitation.

The ability of POPs to travel great distances is part of the explanation for why countries that banned the use of specific POPs are no longer experiencing a decline in their concentrations; the wind may carry chemicals into the country from places that still use them.

Health Effects

POP exposure can cause death and illnesses including disruption of the endocrine, reproductive, and immune systems; neurobehavioural disorders; and cancers possibly including breast cancer. Exposure to POPs can take place through diet, environmental exposure, or accidents. Knowledge of the bioavailability of the pollutant as well as the hazard and dose/response relationship is essential in performing a human health risk assessment.

A study published in 2006 indicated a link between blood serum levels of POPs and diabetes. Individuals with elevated levels of persistent organic pollutants (DDT, dioxins, PCBs and Chlordane, among others) in their body were found to be up to 38 times more likely to be insulin resistant than individuals with low levels of these pollutants, though the study did not demonstrate a cause and effect relationship. As most exposure to POPs is through consumption of animal fats, study participants with high levels of serum POPs are also very likely to be consumers of high amounts of animal fats, and thus the consumption of the fats themselves, or other associated factors may be responsible for the observed increase in insulin resistance. Another possibility is that insulin resistance causes increased accumulation of POPs. Among study participants, obesity was associated with diabetes only in people who tested high for these pollutants. These pollutants are accumulated in animal fats, so minimizing consumption of animal fats may reduce the risk of diabetes. According to the US Department of Veterans Affairs, type 2 diabetes is on the list of presumptive diseases associated with exposure to

Agent Orange (which contained 2,3,7,8-tetrachlorodibenzodioxin) in the Vietnam War.

Environmental Persistent Pharmaceutical Pollutants EPPP

The term Environmental Pharmaceutical Persistent Pollutant (EPPP), was suggested in the nomination 2010 of pharmaceuticals and environment as an emerging issue to Strategic Approach to International Chemicals Management (SAICM) by the International Society of Doctors for the Environment (ISDE). Pharmaceuticals are synthetic chemicals belonging to a wide group of different chemical families and may also react different in the environment.

There exists very well documented evidence that some pharmaceuticals enter and persist in the environment, some are endocrine disruptors (synthetic hormones), some are designed to kill bacteria and viruses (antibiotics) and may affect microorganism and wild life in severe and unexpected ways. Little is known on the possible negative effects and impacts of EPPP in humans and the environment by diffuse and systematic exposure, for long periods of time, especially during the vulnerable periods of development.

As there are thousands of different synthesized chemicals present at the same time in the environment, different interactions may occur and the result of these multiple exposure in human and nature are not sufficiently studied or understood. EPPPs are insufficiently addressed as not covered by other international or regional agreements or arrangements. Pharmaceuticals comprise one of the few groups of chemicals specifically designed to act on living cells, which presents a special risk when they enter, persist and disseminate in the environment. Pharmaceutical chemicals are designed to be non-degradable to resist the acid environment in the stomach, and to be long-lasting; to be administrated according to a specific defined time schedule. Pharmaceuticals chemicals, widely used globally by humans and for food production for an intended purpose, may enter and persist in the environment during their life cycle creating a new and emerging problem, and may pose a threat of important magnitude, with significant adverse effects on environment and human health and special impact in vulnerable populations.

As the world's population is growing and ageing, more people can afford medical treatment and new treatments are developed, the amounts of pharmaceuticals can be expected to increase rapidly. Pharmaceuticals chemicals entering the environment persist there and residues are presently found in drinking water. They are found in fish where they may accumulate. The presence of different pharmaceutical chemicals contributes to the increasing multiple chemical cocktail that today's population is exposed to. Vulnerable populations are exposed, for example foetuses during the windows of development, with possible important consequences for life. With exception for downstream sewage plants, the concentration of pharmaceuticals in water is probably extremely low. However, the effect that the chronic exposure to environmental pharmaceuticals chemicals adds to the effects of other chemicals in the cocktail is still not studied. The different chemicals might be potentiating synergistic effects (1+1=3). An extremely sensitive group in this respect are foetuses.

EPPPs are already found in water all over the world. The diffuse exposure might contribute to

- extinction of species and imbalance of sensible ecosystems, as many EPPPs affect the reproductive systems of for example frogs, fish and mussels;
- genetic, developmental, immune and hormonal health effects to humans and other species, in the same way as e.g. oestrogen-like chemicals;
- development of microbes resistant to antibiotics, as is found in India.

Release

Pharmaceuticals reach the environment mainly in three ways:

- They are excreted from humans and animals, intact or metabolized, mainly into the urine, passing on to the environment directly or via sewage plants.
- Unused reach the environment either via household water or via urban solid garbage handling.
- Manufacturing plants producing the active substances might unintentionally release pharmaceuticals into the environment.

Some pharmaceuticals are degraded to various extents in sewage treatment plants, but others leave the plant in active forms. Active residues of pharmaceuticals have been detected in surface water, and they may persist in the environment for long periods of time. Large amounts of antibiotics and other pharmaceuticals have been found downstream from sewage plants for pharmaceutical industries. EPPPs from sewage sludge used as fertilizer are absorbed by soya, and antibiotics have been found in the leaves.

Drinking Water

Which EPPPs are found in drinking water depends on what resources and detection methods are available. Atenolol (beta blocker), citalopram (antidepressant drug), diclofenak (analgesic), ibuprofen (analgesic), metoprolol (beta blocker), naproxen (anti-inflammatory) and trimetoprim (antibiotic) have been found in drinking water of Stockholm, Sweden. Fish caught downstream from the sewage plants of Stockholm contain EPPPs like citalopram (antidepressant drug) and propoxyphene (narcotic/anesthetic).

Several broad-spectrum antibiotics in very high concentrations, as well as bacteria resistant to all known antibiotics, were found downstream from a sewage plant in India. Also in Indian drinking water cetirizin (antihistaminic), ciprofloxacin (antibiotic), enoxacin (antibiotic), terbinafin (antimycotic), and citalopram (antidepressant drug) were found. Up to 14 different pharmaceuticals have been found in the drinking water of big cities around the world. There also exist publications reporting the presence of cancer drugs in surface water in some countries. Some of these environmental pharmaceuticals chemicals are well known to have serious genotoxic effects in humans. Many are not very well studied for their toxic effects on human periods of development. Half-life in nature varies depending on the environment (air, water, soil, sludge), but is more than one year for several compounds. Clofibric acid, a metabolite of the lipid-lowering agent clofibrate, can still be found in surface as well as well water, although clofibrate long ago has been withdrawn. Concentrations of EPPPs can vary from 1 ng to 1 mg per litre (2). Serious effects of EPPPs on water-living organisms, especially on reproductive systems, have been already shown, as well as on microbial communities.

Exposure

Concentrations in surface waters, groundwater and partially treated water are typically less than 0.1 µg/l (or 100 ng/l), and concentrations in treated water are generally below 0.05 µg/l (or 50 ng/l).(ny 8 WHO) However, all water on the earth is part of the same stable pool, and as larger amounts of pharmaceuticals are consumed, there is a risk that the concentration of pharmaceuticals in drinking water will increase. The tendency of bio-accumulation in fish is alarming, as fish is an important nourishment. The impact of pharmaceutical chemicals, due to diffuse exposure by their presence in the water environment, might contribute to the wide chemical exposure of all species and to their possible extinction, as well as to the imbalance in sensitive eco-systems. Consequences for human health and the equilibrium of the biological environmental system may be irreversible. Multiple human exposures to EPPP may start at conception and may be combined with a cocktail of other chemicals present in the environment. The effects of exposure to these mixtures are difficult to understand due the complexity of the situation during a period of special vulnerability and sensitivity, but can not be denied. Another very serious threat is development and spread of bacteria, viruses and other microbes resistant to the antibiotics present in the environment, with possible unpredictable important consequences.

Assessment

- Pharmaceuticals are special kinds of chemicals. They are manufactured to be biologically active in living organisms, to be persistent to biodegradation and to have long half-lives. This makes them more risky in nature. Release is ongoing always and everywhere, diffuse and impossible to control. They cannot be forbidden.
- The levels of pharmaceuticals in surface or drinking water are generally below 1 mg per litre, often measured in ng per litre (2, 8). This low concentration might appear to guarantee that they hardly pose any problem to public health. Assuming a concentration of 100 ng/l of a pharmaceutical that in humans has DDD (defined daily dose) of 10 mg implies that a volume of 100,000 litres would be required to make up one single DDD. Such calculation, however, is an over-simplification that does not take into account several important dynamic aspects of the low chronic exposure to concentrations of pharmaceuticals in the water or the vulnerable population exposure for example during the period of development.
- Aquatic organisms may bio-concentrate and bio-accumulate lipid soluble chemicals, including pharmaceuticals. It is well known that certain fish species, like herring, may contain very high concentrations of the persistent and lipophilic chemicals DDT (dichlorodiphenyl-trichloroethane, an insecticide) and PCB (polychlorinated biphenils, a group of industry chemicals earlier used in e.g. building materials). The same mechanism may also be applied for chemicals synthesized for pharmaceutical uses. Bioackumulation of citalopram (SSRI, antidepresssivum) and propoxyfen (painkiller) has been found in perche in the

Baltic Sea. Therapeutic levels of levonorgestrel (a sex hormone) has been found in Rainbow trout downstream a sewage plant.

- Pharmaceutical chemicals are not thought or designed to enter in the environment and persist there but for a clear pharmaceutical purpose. Pharmaceutical are synthetic chemicals, they belong to a wide group of different chemical families and may also react differently in the environment. When a new medicine is developed, it's pharmacological and toxicological effects is tested in acute trials, before being accepted for marketing. However, clinical test procedures are not entirely sufficient to completely guarantee that a new pharmaceutical is devoid of unacceptable side effects when used in large cohorts of patients for a long time. Furthermore, there are currently no test methods to assess whether such effects may occur after long-term use in human, during periods of development, on aquatic microorganisms or how they may affect other animals. Based on this, the persistent and diffuse exposure to low doses of pharmaceutical synthetic chemicals, for long periods of time, is not currently well known or studied.
- The diffuse dissemination of the EPPP in the environment may indiscriminately expose vulnerable populations: embryos/ foetuses, children and adolescents, men and women of reproductive age, and elderly or weak persons. Some of the pharmaceuticals found in surface water are prescribed to patients under special controlled conditions for short periods of time due the risk of side effects. Others are prohibited from prescription to pregnant women or children. These chemicals were not synthesized to expose the general population in a diffuse manner. This presents a new and emerging issue under the chemical safety global pollution.
- It can be assumed that a large portion of excreted or disposed medicines reach the public sewage treatment plants (STP's) . Today, the sewage plants do not have the capacity to clean the water from pharmaceutical chemicals. This is sometimes also the case for the industries' own sewage plants. In many parts of the world, the sewage plant water is reused as drinking water, not always after cleaning treatment. To add a step for cleaning sewage water from pharmaceuticals means more energy, more chemicals and higher costs. - Alternatively, the sewage is directly let out into various surface waters like

rivers, lakes, streams or the open sea. - Detection and monitoring at global scale of EPPPs in drinking and surface water as in animals and plants is necessary to understand the magnitude of the problem. The first step is to recognize EPPP as an emerging issue to be able to invest the necessary human and financial resources and develop effective environmental detection methods.

Laws and Regulations

Environmental Pharmaceutical Persistent Pollutants, EPPP, are insufficiently addressed as not covered by other international or regional agreements or arrangements. Pharmaceuticals differ from other anthropogenic chemicals with respect to legal requirements. They are regularly excluded in laws and regulations which control manufacture, marketing, use, and disposal of other consumer products of a chemical character (solvents, paints, glues etc.). As a consequence the possible negative environmental impact of pharmaceuticals is much less documented, in comparison to other consumer chemicals.

In the European Union, the new directive for human pharmaceuticals explicitly requires that all member states should establish collection systems for unused or expired medicines. Such systems were already in use in several member countries at the time the new legislation went into action in 2004.

Nevertheless, the extent to which such systems have been established and made publicly known, varies between regions. Furthermore, the directive does not regulate how the collected pharmaceuticals should be handled. Disposal into the sewage system is still a legally accepted route of elimination. However, incineration at high temperature (1200°C) is a preferred alternative to avoid environmental pollution.

For pharmaceuticals approved for marketing in EU before 1995, there are no requirements for documentation of environmental effects. Hence, pharmaceuticals which have been on the market for decades may have serious environmental effects that have not been detected.

Known Effects of Pharmaceuticals in the Environment

Estradiol (Estrogen, Synthetic Hormone): Concentrations in surface water alone are not sufficient to assess the risk of negative environmental effects in the aquatic environment. Synthetic hormones are endocrine disruptors. Thus, estrogenic compounds like ethinylestradiol (estrogen hormone) at concentrations < 1 ng per litre may cause both vitellogenin production (a frequently used index for feminization of male fish), and structural change in their sex organs. It has also been demonstrated that fish exposed to sewage treatment plant (STP) effluent can take up and concentrate estrogenic compounds, including ethinyl-estradiol, to very high internal levels. These observations on feminization of fish by estrogenic compounds in STP effluents have been observed in many countries, and have also been observed in other species, like frogs, alligators and molluscs.

Cardiovascular Medicines

Other examples of environmental impact in the aquatic environment of human medication concern both cardiovascular and neuro-psychiatric medicines. The non-selective beta-blocking agent propanolol was found to cause a significant decrease in egg production in medaka fish, at a concentration close to that demonstrated in the sewage treatment plants (STP) effluents. Gemfibrozil (cholesterol and triglycerides lowering drug) often appears in the effluent from STPs. At concentrations close to those reported in STP effluent, gemfibrozil lowers the blood levels of testosterone in fish.

Citalopram / Fluoxetine (serotonin reuptake inhibitor anti depressants, SSRI's) Some SSRI's have been shown to accumulate in exposed fish. Citalopram has been detected in liver from wild perch in low μg per kg levels, and fluoxetine affects the serotonin system in the same way that it does in humans. Fluoxetine has also been shown to affect swimming activity in shellfish; whether this is linked to a disturbance of serotonin function in the brain is still unknown.

Antibiotics

High levels of antibiotics in the water are a cause for alarm as there is an increased risk of selecting resistant bacteria, an issue of global concern. This can lead to some highly effective antibiotics becoming ineffective. There are several examples. In India, bacteria resistant to ciprofloxacin have been found downstream pharmaceutical plants, genes for multi resistance were found in drinking water, and multi resistant Salmonella in water sprayed on vegetables. From Europe we know about the epidemic with multi resistant EHEC in summer 2011, originating from water sprayed vegetables. The term "eco-shadow" has been introduced to describe the ecological impact of antibiotics. Antibiotics with a wide spectrum that are also stable will have a greater impact on the bacterial flora (a long eco-shadow) than those with a narrow antibacterial spectrum which disintegrates more rapidly (a short eco-shadow).

The ecological effects of tetracyclines and quinolones have been observed. They are not metabolized in the human body and are therefore excreted unmodified. When entered into the environment they are poorly degraded. They can be toxic to other animals, affecting particularly microorganism and fish. In the effluent from a sewage plant in India, several broad spectrum antibiotics were found in concentrations toxic to bacteria and plants. In the sewage plant itself, there were enterococcae resistant to all known antibiotics. The development of resistant bacteria in sewage plants is stimulated by high concentration of antibiotics (e.g. in plant sewage), large amounts of bacteria (e.g. from human sewage water that is added in plant sewage), and selection of Information that can be used to assess the nominated issue have been observed.

Possibilities

Environmental Classification of Pharmaceuticals: In Sweden, the industry together with universities and health care sector has developed a method for environmental risk assessment and environmental classification of drugs. Environmental risk refers to the risk of toxicity to the aquatic environment. It is based on the ratio between predicted environmental concentration of the substance (PEC) and the highest concentration of the substance that does not have a harmful effect in the environment (PNEC). Environmental hazard expresses the inherent environmentally damaging characteristics of the substance in terms of persistence, bioaccumulation and toxicity. The toxicity tests used are acute toxicity of fish, acute toxicity of Daphnia sp. and growth inhibition test of algae. Most medications on the Swedish market are now classified. This gives the health care possibilities to make better choices when prescribing medicines.

Good Manufacturing Practice

"Good manufacturing practice" or "GMP" are practices and the systems required to be adapted in pharmaceutical manufacturing, quality control, quality system covering the manufacture and testing of pharmaceuticals or drugs including active pharmaceutical ingredients, diagnostics, foods, pharmaceutical products, and medical devices. The World Health Organization (WHO) version of GMP is used by pharmaceutical regulators and the pharmaceutical industry in over one hundred countries worldwide, primarily in the developing world. So far, emissions into the environment are not included.

Gaps in Knowledge

Effective environmental detection methods have to be developed and global detection strategy applied to map the current global situation. There are currently no test methods to assess whether negative effects may occur after long-term environmental diffuse exposure in humans, during the vulnerable periods of development, on aquatic micro-organism or how it may affect other animals. Therefore the precautionary principle must be guiding. Concentrations in surface water alone are not sufficient to assess the risk of negative environmental effects of these synthetic chemicals. Consideration must be taken to bio-accumulation in fish and other aquatic food used by humans, as well as to additive and synergetic effects between pharmaceutical and other chemicals in the contaminated water.

In a small study, several pharmaceuticals were found in milk of goat, cow and human. More research is needed to find out how common this is, the concentrations and the sources. The industry must be invited to actively work on reducing pharmaceuticals in the environment. Emission of pharmaceuticals should be included in GMP.

Polycyclic Aromatic Hydrocarbons

Polycyclic aromatic hydrocarbons (PAHs), also known as poly-aromatic hydrocarbons or polynuclear aromatic hydrocarbons, are potent atmospheric pollutants that consist of fused aromatic rings and do not contain heteroatoms or carry substituents. Naphthalene is the simplest example of a PAH. PAHs occur in oil, coal, and tar deposits, and are produced as byproducts of fuel burning (whether fossil fuel or biomass). As a pollutant, they are of concern because some compounds have been identified as carcinogenic, mutagenic, and teratogenic. PAHs are also found in cooked foods. Studies have shown that high levels of PAHs are found, for example, in meat cooked at high temperatures such as grilling or barbecuing, and in smoked fish. They are also found in the interstellar medium, in comets, and in meteorites and are a candidate molecule to act as a basis for the earliest forms of life.

Occurrence and Pollution

Polycyclic aromatic hydrocarbons are lipophilic, meaning they mix more easily with oil than water. The larger compounds are less water-soluble and less volatile. Because of these properties, PAHs in the environment are found primarily in soil, sediment, and oily substances, as opposed to in water or air. However, they are also a component of concern in particulate matter suspended in air.

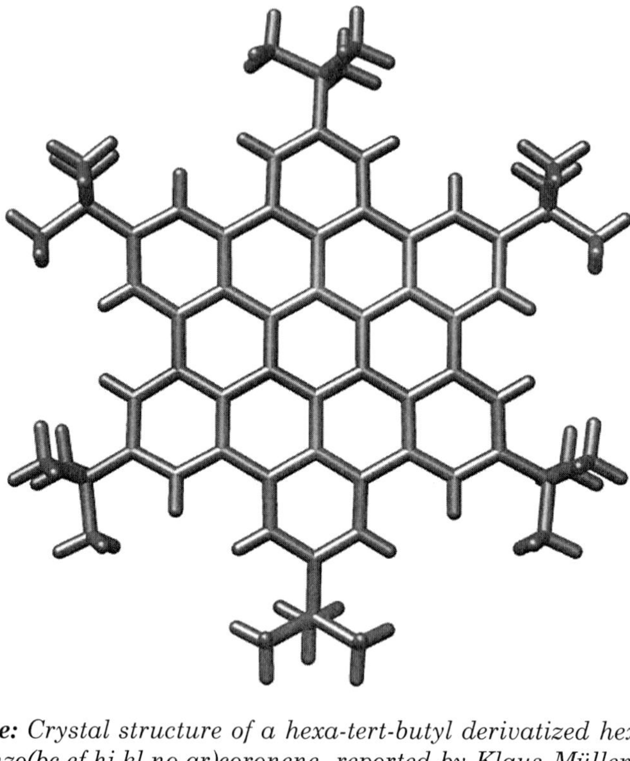

Figure: *Crystal structure of a hexa-tert-butyl derivatized hexa-peri-hexabenzo(bc,ef,hi,kl,no,qr)coronene, reported by Klaus Müllen and co-workers. The tert-butyl groups make this compound soluble in common solvents such as hexane, in which the unsubstituted PAH is insoluble.*

Natural crude oil and coal deposits contain significant amounts of PAHs, arising from chemical conversion of natural product molecules, such as steroids, to aromatic hydrocarbons. They are also found in processed fossil fuels, tar and various edible oils. In a study evaluating the genotoxic and carcinogenic risks associated with the consumption of repeatedly heated coconut oil (RCO), one of the commonly consumed cooking and frying medium, it was concluded that dietary consumption of RCO can cause a genotoxic and preneoplastic change in the liver.

PAHs are one of the most widespread organic pollutants. In addition to their presence in fossil fuels they are also formed by incomplete combustion of carbon-containing fuels such as wood, coal, diesel, fat, tobacco, and incense. Different types of combustion yield different distributions of PAHs in both relative amounts of individual PAHs and in which isomers are produced. Thus, coal burning produces a different mixture than motor-fuel combustion or a forest fire, making the compounds potentially useful as indicators of the burning history. Hydrocarbon emissions from fossil fuel-burning engines are regulated in developed countries.

PAH chemicals are often linked to oil spills. Following the massive Deepwater Horizon oil spill, scientists found PAH levels to be 40 times higher than before the area was affected. The researchers said that the compounds can enter the food chain through organisms such as plankton or fish.

Human Health

The toxicity of PAHs is structure-dependent. Isomers (PAHs with the same formula and number of rings) can vary from being nontoxic to extremely toxic. One PAH compound, benzo[*a*]pyrene, is notable for being the first chemical carcinogen to be discovered (and is one of many carcinogens found in cigarette smoke). The EPA has classified seven PAH compounds as probable human carcinogens: benz[*a*]anthracene, benzo[*a*]pyrene, benzo[*b*]fluoranthene, benzo [*k*] fluoranthene, chrysene, dibenz(a,h)anthracene, and indeno(1,2,3-cd) pyrene.

PAHs known for their carcinogenic, mutagenic, and teratogenic properties are benz[*a*]anthracene and chrysene, benzo[*b*]fluoranthene, benzo[*j*]fluoranthene, benzo[*k*]fluoranthene, benzo[*a*]pyrene, benzo [*ghi*] perylene, coronene, dibenz(a,h)anthracene ($C_{20}H_{14}$), indeno(1,2,3-cd)pyrene ($C_{22}H_{12}$), and ovalene.

High prenatal exposure to PAH is associated with lower IQ and childhood asthma. The Center for Children's Environmental Health reports studies that demonstrate that exposure to PAH pollution during pregnancy is related to adverse birth outcomes including low birth weight, premature delivery, and heart malformations. Cord blood of exposed babies shows DNA damage that has been linked to cancer. Follow-up studies show a higher level of developmental delays at age three, lower scores on IQ tests and increased behaviourial problems at ages six and eight.

In addition, a 2012 Columbia University study in *Environmental Health Perspectives* linked prenatal exposure to pollutants and eventual child behavioural outcomes.

The study found that exposure to higher levels of PAH was associated with a 24% higher score of anxiety/depression for children ages 6 to 7 than those with low exposure levels. Infants found to have elevated PAH levels in their umbilical cord blood were 46% more likely to eventually score highly on the anxiety/depression scale than those with low PAH levels in cord blood.

List of PAHs

Although the health effects of individual PAHs are not exactly alike, the following 18 PAHs are considered as a group in this profile issued by the Agency for Toxic Substances and Disease Registry(ATSDR):

- acenaphthene
- acenaphthylene
- anthracene
- benz[*a*]anthracene
- benzo[*a*]pyrene
- benzo[*e*]pyrene
- benzo[*b*]fluoranthene
- benzo[*ghi*]perylene
- benzo[*j*]fluoranthene
- benzo[*k*]fluoranthene
- chrysene
- coronene
- dibenz(a,h)anthracene
- fluoranthene
- fluorene
- indeno(1,2,3-cd)pyrene
- phenanthrene
- pyrene

Additional PAHs:

- naphthalene

Chemistry

The simplest PAHs, as defined by the International Union of Pure and Applied Chemistry (IUPAC) (G.P Moss, IUPAC nomenclature for fused-ring systems), are phenanthrene and anthracene, which both contain three fused aromatic rings. Smaller molecules, such as benzene, are not PAHs. PAHs may contain four-, five-, six- or seven-member rings, but those with five or six are most common. PAHs composed only of six-membered rings are called alternant PAHs.

Certain alternant PAHs are called benzenoid PAHs. The name comes from benzene, an aromatic hydrocarbon with a single, six-membered ring. These can be benzene rings interconnected with each other by single carbon-carbon bonds and with no rings remaining that do not contain a complete benzene ring. The set of alternant PAHs is closely related to a set of mathematical entities called polyhexes, which are planar figures composed by conjoining regular hexagons of identical size.

PAHs containing up to six fused aromatic rings are often known as "small" PAHs, and those containing more than six aromatic rings are called "large" PAHs. Due to the availability of samples of the various small PAHs, the bulk of research on PAHs has been of those of up to six rings. The biological activity and occurrence of the large PAHs does appear to be a continuation of the small PAHs.

They are found as combustion products, but at lower levels than the small PAHs due to the kinetic limitation of their production through addition of successive rings. In addition, with many more isomers possible for larger PAHs, the occurrence of specific structures is much smaller.

PAHs possess very characteristic UV absorbance spectra. These often possess many absorbance bands and are unique for each ring structure. Thus, for a set of isomers, each isomer has a different UV absorbance spectrum than the others. This is particularly useful in the identification of PAHs. Most PAHs are also fluorescent, emitting characteristic wavelengths of light when they are excited (when the molecules absorb light). The extended pi-electron electronic structures of PAHs lead to these spectra, as well as to certain large PAHs also exhibiting semi-conducting and other behaviours.

Naphthalene ($C_{10}H_8$, a constituent of some mothballs), consisting of two coplanar six-membered rings sharing an edge, is another aromatic hydrocarbon. By formal convention, it is not a true PAH though is referred to as a bicyclic aromatic hydrocarbon. Aqueous solubility decreases approximately one order of magnitude for each additional ring.

Aromaticity

Although PAHs clearly are aromatic compounds, the degree of aromaticity can be different for each ring segment. According to Clar's rule (formulated by Erich Clar in 1964) for PAHs the resonance structure with the most disjoint aromatic ï-sextets—i.e. benzene-like moieties—is the most important for the characterization of the properties.

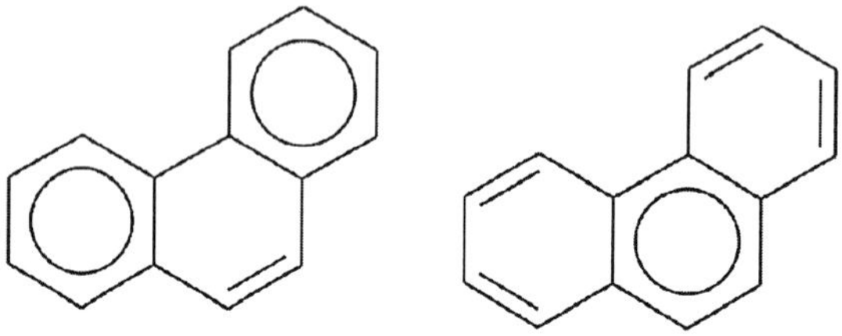

Figure: *Phenanthrene (1)*

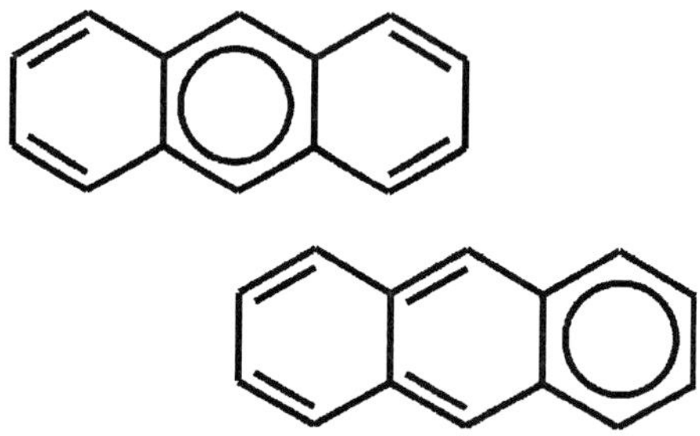

Figure: *Anthracene (2)*

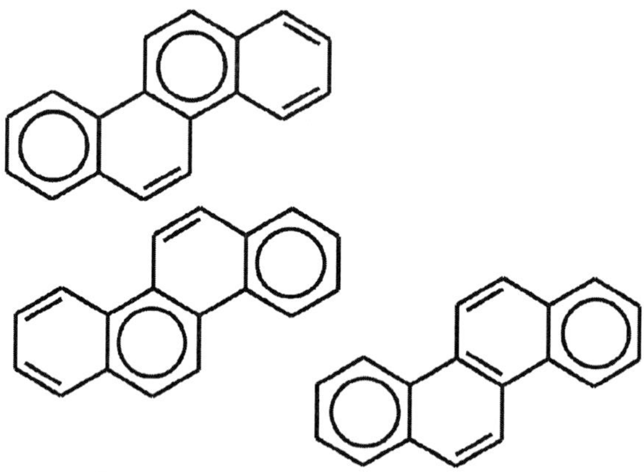

Figure: *Chrysene (3)*

For example, in phenanthrene (1) one Clar structure has two sextets at the extremities, while the other resonance structure has just one central sextet. Therefore in this molecule the outer rings are firmly aromatic while its central ring is less aromatic and therefore more reactive.

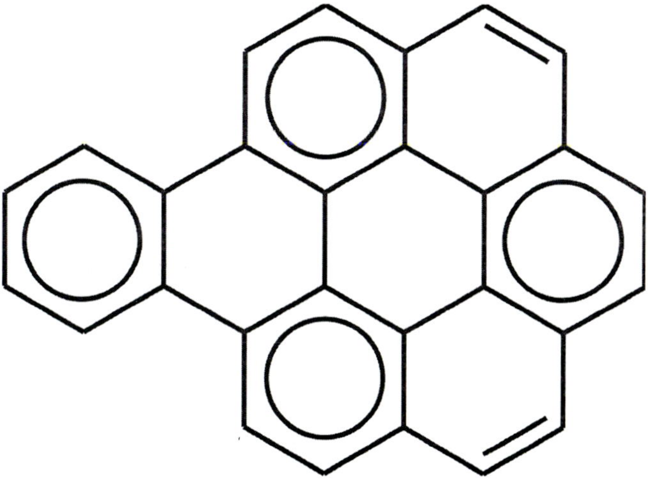

Figure: *Clar rule*

In contrast, in anthracene (2) the number of sextets is just one and aromaticity spreads out. This difference in number of sextets is reflected in the UV absorbance spectra of these two isomers. Phenanthrene has a highest wavelength absorbance around 290 nm, while anthracene has highest wavelength bands around 380 nm. Three Clar structures with two sextets are present in chrysene (3) and by superposition the aromaticity in the outer ring is larger than in the inner rings. Another relevant Clar hydrocarbon is zethrene.

Origins of Life

In January 2004 (at the 203rd Meeting of the American Astronomical Society), it was reported that a team led by A. Witt of the University of Toledo, Ohio studied ultraviolet light emitted by the Red Rectangle nebula and found the spectral signatures of anthracene and pyrene (no other such complex molecules had ever before been found in space). This discovery was considered as a controversial confirmation of a hypothesis that as nebulae of the same type as the Red Rectangle approach the ends of their lives, convection currents cause carbon and hydrogen in the nebulae's core to get caught in stellar winds, and radiate outward. As they cool, the atoms supposedly bond to each other in various ways and eventually form particles of a million or more atoms. Witt and his team inferred that since they discovered PAHs—which may have been vital in the formation of early life on Earth—in a nebula, by necessity they must originate in nebulae. More recently, fullerenes (or "buckyballs"), have been detected in other nebulae. Fullerenes are also implicated in the origin of life;

according to astronomer Letizia Stanghellini, "It's possible that buckyballs from outer space provided seeds for life on Earth." In September 2012, NASA scientists reported that PAHs, subjected to interstellar medium (ISM) conditions, are transformed, through hydrogenation, oxygenation, and hydroxylation, to more complex organics—"a step along the path toward amino acids and nucleotides, the raw materials of proteins and DNA, respectively". Further, as a result of these transformations, the PAHs lose their spectroscopic signature which could be one of the reasons "for the lack of PAH detection in interstellar ice grains, particularly the outer regions of cold, dense clouds or the upper molecular layers of protoplanetary disks."

Volatile Organic Compounds

Volatile organic compounds (VOCs) are organic chemicals that have a high vapour pressure at ordinary, room-temperature conditions. Their high vapour pressure results from a low boiling point, which causes large numbers of molecules to evaporate or sublimate from the liquid or solid form of the compound and enter the surrounding air. An example is formaldehyde, with a boiling point of –19 °C (–2 °F), slowly exiting paint and getting into the air. VOCs are numerous, varied, and ubiquitous. They include both human-made and naturally occurring chemical compounds. Most scents or odours are of VOCs. VOCs play an important role in communication between plants. Some VOCs are dangerous to human health or cause harm to the environment. Anthropogenic VOCs are regulated by law, especially indoors, where concentrations are the highest. Harmful VOCs are typically not acutely toxic, but instead have compounding long-term health effects. Because the concentrations are usually low and the symptoms slow to develop, research into VOCs and their effects is difficult.

Definitions

Diverse definitions of the term VOC are in use. The definitions of VOCs used for control of precursors of photochemical smog used by EPA, and U.S. States with their own outdoor air pollution regulations include exemptions for VOCs that are determined to be non-reactive, or of low-reactivity in the smog formation process. EPA formerly defined these compounds as reactive organic gases (ROG) but changed the terminology to VOC.

In the USA, different regulation exists per state - most prominent is the VOC regulation by SCAQMD and by California Air Resources

Board. However, this specific use of the term VOCs can be misleading, specifically when applied to indoor air quality because many chemicals that are not regulated for controlling outdoor air pollution can still be important for indoor air pollution - there is no correlation between VOC content in a product and VOC emissions from that product into indoor air.

Canada

Health Canada classes VOCs as organic compounds that have boiling points roughly in the range of 50 to 250 °C (122 to 482 °F). The emphasis is placed on commonly encountered VOCs that would have an effect on air quality.

European Union

A VOC is any organic compound having an initial boiling point less than or equal to 250 °C (482 °F) measured at a standard atmospheric pressure of 101.3 kPa and can do damage to visual or audible senses.

US

VOCs (or specific subsets of the VOCs) are legally defined in the various laws and codes under which they are regulated. Other definitions may be found from government agencies investigating or advising about VOCs. The United States Environmental Protection Agency (EPA) regulates VOCs in the air, water, and land. The Safe Drinking Water Act implementation includes a list labelled "VOCs in connection with contaminants that are organic and volatile." The EPA also publishes testing methods for chemical compounds, some of which refer to VOCs.

In addition to drinking water, VOCs are regulated in discharges to waters (sewage treatment and stormwater disposal), as hazardous waste, but not in non industrial indoor air. The United States Department of Labor and its Occupational Safety and Health Administration (OSHA) regulate VOC exposure in the workplace. Volatile organic compounds that are hazardous material would be regulated by the Pipeline and Hazardous Materials Safety Administration while being transported.

Biologically Generated VOCs

Not counting methane, biological sources emit an estimated 1150 teragrams of carbon per year in the form of VOCs. The majority of VOCs are produced by plants, the main compound being isoprene. The remainder are produced by animals, microbes, and fungi, such as molds. The strong odor emitted by many plants consists of green leaf volatiles, a subset of VOCs. Emissions are affected by a variety of factors, such as temperature, which determines rates of volatilization and growth, and sunlight, which determines rates of biosynthesis. Emission occurs almost exclusively from the leaves, the stomata in particular. A major class of VOCs is terpenes, such as myrcene. Providing a sense of scale, a forest 62,000 km^2 in area (the U.S. state of Pennsylvania) is estimated to emit 3,400,000 kilograms of terpenes on a typical August day during the growing season. Induction of genes producing volatile organic compounds, and subsequent increase in volatile terpenes has been achieved in maize using (Z)-3-Hexen-1-ol and other plant hormones.

Anthropogenic Sources

Anthropogenic sources emit about 142 teragrams of carbon per year in the form of VOCs.

Specific Components

Paints and Coatings: A major source of man-made VOCs are coatings, especially paints and protective coatings. Solvents are required to spread a protective or decorative film. Approximately 12 billion litres of paints are produced annually. Typical solvents are aliphatic hydrocarbons, ethyl acetate, glycol ethers, and acetone. Motivated by cost, environmental concerns, and regulation, the paint and coating industries are increasingly shifting toward aqueous solvents.

Chlorofluorocarbons and Chlorocarbons

Chlorofluorocarbons, which are banned or highly regulated, were widely used cleaning products and refrigerants. Tetrachloroethene is used widely in dry cleaning and by industry. Industrial use of fossil fuels produces VOCs either directly as products (e.g., gasoline) or indirectly as byproducts (e.g., automobile exhaust).

Benzene

One VOC that is a known human carcinogen is benzene, which is a chemical found in environmental tobacco smoke, stored fuels, and exhaust from cars in an attached garage. Benzene also has natural sources such as volcanoes and forest fires. It is frequently used to make other chemicals in the production of plastics, resins, and synthetic fibres. Benzene evaporates into the air quickly and the vapor of benzene is heavier than air allowing the compound to sink into low-lying areas. Benzene has also been known to contaminate food and water and if digested can lead to vomiting, dizziness, sleepiness, rapid heartbeat, and at high levels, even death may occur.

Methylene Chloride

Methylene chloride is another VOC that is highly dangerous to human health. It can be found in adhesive removers and aerosol spray paints and the chemical has been proven to cause cancer in animals. In the human body, methylene chloride is converted to carbon monoxide and a person will suffer the same symptoms as exposure to carbon monoxide. If a product that contains methylene chloride needs to be used the best way to protect human health is to use the product outdoors. If it must be used indoors, proper ventilation is essential to keeping exposure levels down.

Perchloroethylene

Perchloroethylene is a volatile organic compound that has been linked to causing cancer in animals. It is also suspected to cause many of the breathing related symptoms of exposure to VOCs. Perchloroethylene is used mostly in dry cleaning. While dry cleaners recapture perchloroethylene in the dry cleaning process to reuse it, some environmental release is unavoidable. Studies show that people breathe in low levels of this VOC in homes where dry-cleaned clothes are stored and while wearing dry-cleaned clothing.

MTBE

MTBE was banned in the US around 2004 in order to limit further contamination of drinking water aquifers primarily from leaking underground gasoline storage tanks where MTBE was used as an octane booster and oxygenated-additive.

Indoor air

Since people today spend most of their time at home or in an office, long-term exposure to VOCs in the indoor environment can contribute to sick building syndrome. In offices, VOC results from new furnishings, wall coverings, and office equipment such as photocopy machines, which can off-gas VOCs into the air. Good ventilation and air-conditioning systems are helpful at reducing VOC emissions in the indoor environment. Studies also show that relative leukemia and lymphoma can increase through prolonged exposure of VOCs in the indoor environment. There are two standardized methods for measuring

VOCs, one by the National Institute for Occupational Safety and Health (NIOSH) and another by Occupational Safety and Health Administration (OSHA). Each method uses a single component solvent; butanol and hexane cannot be sampled, however, on the same sample matrix using the NIOSH or OSHA method. The aromatic VOC compound benzene, emitted from exhaled cigarette smoke is labelled as carcinogenic, and is ten times higher in smokers than in nonsmokers.

The United States Environmental Protection Agency (EPA) has found concentrations of VOCs in indoor air to be 2 to 5 times greater than in outdoor air and sometimes far greater. During certain activities indoor levels of VOCs may reach 1,000 times that of the outside air. Studies have shown that individual VOC emissions by themselves are not that high in an indoor environment, but the indoor total VOC (TVOC) concentrations can be up to five times higher than the VOC outdoor levels. New buildings especially, contribute to the highest level of VOC off-gassing in an indoor environment because of the abundant new materials generating VOC particles at the same time in such a short time period. In addition to new buildings, we also use many consumer products that emit VOC compounds, therefore the total concentration of VOC levels is much greater within the indoor environment. VOC concentration in an indoor environment during winter is three to four times higher than the VOC concentrations during the summer. High indoor VOC levels are attributed to the low rates of air exchange between the indoor and outdoor environment as a result of tight-shut windows and the increasing use of humidifiers.

Regulation of Indoor VOC Emissions

In most countries, a separate definition of VOCs is used with regard to indoor air quality that comprises each organic chemical compound that can be measured as follows: Adsorption from air on Tenax TA, thermal desorption, gas chromatographic separation over a 100% nonpolar column (dimethylpolysiloxane). VOC (volatile organic compounds) are all compounds that appear in the gas chromatogram between and including n-hexane and n-hexadecane. Compounds appearing earlier are called VVOC (very volatile organic compounds) compounds appearing later are called SVOC (semi-volatile organic compounds). France and Germany have enacted regulations to limit VOC emissions from commercial products, and industry has developed numerous voluntary ecolabels and rating systems, such as EMICODE, M1, Blue Angel and Indoor Air Comfort In the United States, several standards exist; California Standard CDPH Section 01350 is the most popular one. Over the last few decades, these regulations and standards changed the marketplace, leading to an increasing number of low-emitting products: The leading voluntary labels report that licenses to several hundreds of low-emitting products have been issued.

Formaldehyde

Many building materials such as paints, adhesives, wall boards, and ceiling tiles slowly emit formaldehyde, which irritates the mucous membranes and can make a person irritated and uncomfortable. Formaldehyde emissions from wood are in the range of 0.02 – 0.04 ppm. Relative humidity within an indoor environment can also affect the emissions of formaldehyde. High relative humidity and high temperatures allow more vaporization of formaldehyde from wood-materials.

Health risks

Respiratory, allergic, or immune effects in infants or children are associated with man-made VOCs and other indoor or outdoor air pollutants. Some VOCs, such as styrene and limonene, can react with nitrogen oxides or with ozone to produce new oxidation products and secondary aerosols, which can cause sensory irritation symptoms. Unspecified VOCs are important in the creation of smog.

Health Effects Include: Eye, nose, and throat irritation; headaches, loss of coordination, nausea; damage to liver, kidney, and central nervous system. Some organics can cause cancer in animals; some are suspected or known to cause cancer in humans. Key signs or symptoms associated with exposure to VOCs include conjunctival irritation, nose and throat discomfort, headache, allergic skin reaction, dyspnea, declines in serum cholinesterase levels, nausea, emesis, epistaxis, fatigue, dizziness.

The ability of organic chemicals to cause health effects varies greatly from those that are highly toxic, to those with no known health effects. As with other pollutants, the extent and nature of the health effect will depend on many factors including level of exposure and length of time exposed. Eye and respiratory tract irritation, headaches, dizziness, visual disorders, and memory impairment are among the immediate symptoms that some people have experienced soon after exposure to some organics. At present, not much is known about what health effects occur from the levels of organics usually found in homes. Many organic compounds are known to cause cancer in animals; some are suspected of causing, or are known to cause, cancer in humans.

Reducing Exposure

To reduce exposure to these toxins, one should buy products that contain Low-VOCs or No VOCs. Only the quantity which will soon be needed should be purchased, eliminating stockpiling of these chemicals. Use products with VOCs in well ventilated areas.

When designing homes and buildings, design teams can implement the best possible ventilation plans, call for the best mechanical systems available, and design assemblies to reduce the amount of infiltration into the building.

These methods will help improve indoor air quality, but by themselves they cannot keep a building from becoming an unhealthy place to breathe. While proper building ventilation is a key component to improving indoor air quality, it cannot do the job on its own. As stated earlier, awareness is the key component to improving air quality, when choosing building materials, furnishings, and decorations. When architects and engineers implement best practices in ventilation and mechanical systems, the owner must maintain good air quality levels thereafter.

Limit Values for VOC Emissions

Limit values for VOC emissions into indoor air are published by e.g. AgBB, AFSSET, California Department of Public Health, and others.

Chemical Fingerprinting

The exhaled human breath contains a few hundred volatile organic compounds and is used in breath analysis to serve as a VOC biomarker to test for diseases such as lung cancer. One study has shown that "volatile organic compounds ... are mainly blood borne and therefore enable monitoring of different processes in the body." And it appears that VOC compounds in the body "may be either produced by metabolic processes or inhaled/absorbed from exogenous sources" such as environmental tobacco smoke. Research is still in the process to determine whether VOCs in the body are contributed by cellular processes or by the cancerous tumors in the lung or other organs.

VOCs Sensors

VOCs in the environment or certain atmospheres can be detected based in different principles and interactions between the organic compounds and the sensor components. There are electronic devices that can detect ppm concentrations despite the non-selectivity. Others can predict with reasonable accuracy the molecular structure of the volatile organic compounds in the environment or enclosed atmospheres and could be used as accurate monitors of the Chemical Fingerprint and further as health monitoring devices.

Environmental Xenobiotics

Environmental xenobiotic are xenobiotic substances with a biological activity that are found as pollutants in the natural environment.

Pharmaceuticals: Pharmaceutical drug is a chemical used for the alteration, diagnosis, prevention and treatment of disease, health conditions or structure/function of the human body. Pharmaceutically active compounds (PhACs) are those pharmaceuticals that have by one route or another entered the environment as the parent compound or as pharmacologically active metabolites.

Drugs are developed with the intention of having a beneficial biological effect on the organism to which they are administered, but many such compounds all too often pass into the environment where they may exert an unwanted biological effect.

For many years PhACs have been all but ignored as environmental researchers concentrated on the well known environmentally dangerous chemicals that were/are largely used in agriculture and industry. But with increasing technology to help in the separation and identification of multiple compounds in a mixture, PhACs and their effects have received increasing attention. PhACs have not (until relatively recently) been seen as potentially toxic because regulations associated with pharmaceuticals are typically overseen by human health organizations which have limited experience with environmental issues.

Nearly all categories of pharmaceuticals including pain killers (analgesics and anti inflammatory), antibiotics (antibacterial), anticonvulsant drugs, Beta blockers, blood lipid regulators, X-ray contrast media, cytostatic drugs (Chemotherapy), oral contraceptives, and veterinary pharmaceuticals among many others have been found in the environment.

Sources and Origins

PhACs can be entered into the environment in two main ways; direct and indirect. Indirect sources are PhACs that have performed their biologically intended effect and are passed onto the environment in either their complete or a modified state.

PhAC's can be discharged directly by manufacturers of the pharmaceuticals or effluents from hospitals. However with increasing regulation by local, state and federal regulating agencies, direct discharge is becoming much less of an issue.

There are also several indirect sources of PhACs into the environment. One common indirect source of PhACs into the environment is the passing of antibiotics, anesthetics and growth promoting hormones by domesticated animals in urine and manure. This is often stored in large pits before being pumped and applied to fields as fertilizers where many of the PhACs can be washed away by rainfall to aquatic environments.

Family pets can also be an indirect source of PhACs into the environment.

Most of the PhACs in the environment however come from human sources. A direct human source is leachate from a landfill. Often the pharmaceuticals that are located in landfills are found in their original, most chemically active state.

Most pharmaceuticals are administered and passed through the human body in one of three ways:

1. Metabolized partially or completely within the body and made inactive (Ideal)
2. Partially metabolized and passed through the system
3. Passed through the body unmodified (Worst Case Scenario). In any manner PhACs are then passed to sewage treatment plants (STPs), where facilities are designed to break down natural human waste by microbial degradation. However, many PhACs are of very complex structure and are incompletely broken down in STPs before they are passed into the environment.

Fate in Environment

Once PhACs are entered into the environment they suffer one of three fates:

1. Biodegradation into carbon dioxide and water.
2. Undergo some form of degradation and form metabolites.
3. Persist in the environment unmodified. The amount of the compound that is broken down depends on several factors such as bioavailability and compound structure among many others.

Effects

Because PhACs have come into the limelight relatively recently their effects on the environment are not completely understood. PhACs are also not generally intended to come in contact with the environment, and therefore are not typically tested environmentally prior to release. Therefore several tests are required to determine the different mechanisms and side effects of PhACs in the environment making testing largely impractical.

Many PhACs have very broad modes of action in humans. Similar, subtle reactions may occur in organisms in the environment that are not easily seen by humans. Highly specific mechanisms in humans may solicit profound effects at extremely low concentrations. Many effects may not necessarily be readily detectable and lead to ecological change that would be erroneously attributed to natural change. This said there are several effects that have been identified in the literature.

One long term, possibly irreversible effect is microbiological resistance to antibiotics (antibiotic resistance). Some bacteria may be able to survive when administered antibiotics (especially at low concentrations). Those colonies will multiply and produce new colonies that are resistant to that particular antibiotic and will not succumb the next time antibiotics are administered. Because rivers and streams are ever flowing objects they are an ideal pathway for antibiotics to reach bacteria and therefore provide a source and reservoir for resistant strains to develop and establish themselves.

Another recent discovery is endocrine disruptors. Endocrine disruptors can replace or disturb the balance of hormones within an organism and have been found to be occurring in waters with a concentration in the ng/L level for certain compounds. Some possible effects of endocrine disruptors are male and female sterility, feminization of males, masculinization of females and abnormal testes growth among many others. The exact pathway of occurrence of endocrine disruptors is not completely certain, however several pathways have been proposed.

Typically PhACs are found in low concentrations, (<1 ug/L) making acute toxicity effects fairly unlikely. However, because of their continual input to the environment it is possible for chronic toxicity effects to occur. One major area of concern with several compounds being present at low levels at the same time is what happens when the compounds mix? It is possible and truly likely that these mixtures will have additive, neutralistic or synergistic effects. But again testing would be both time consuming and very expensive to test all of the combined effects.

Common PhACs Found in the Environment

Analgesics (anti-inflammatory and antipyretic) :

1. Acetaminophen
2. Acetylsalicylic Acid
3. Diclofenac
4. Codeine
5. Ibuprofen

Antibiotics

1. Macrolide Antibiotics
2. Sulfonamides
3. Fluoroquinolones
4. Chloraphenicol
5. Tylosin
6. Trimethoprim
7. Erythromycin
8. Lincomycin
9. Sulfamethoxazole
10. Trimethoprim

Anticonvulsant

1. Carbamazepine
2. Primidone

Beta-blockers

1. Metoprolol
2. Propanolol
3. Betaxolol
4. Bisoprolol
5. Nadolol

X-ray Media

1. Iopromide
2. Iopamidol

3. Iohexol
4. Diatrizoate

Cytostatics (Chemotherapy Drugs)

Steroids and hormones :

1. 17α-ethinylestradiol
2. Mestranol
3. 19-norethisterone

Zones of Influence

Pollutants can also be defined by their zones of influence, both horizontally and vertically.

Horizontal Zone: The horizontal zone refers to the area that is damaged by a pollutant. Local pollutants cause damage near the emission source. Regional pollutants cause damage further from the emission source.

Vertical Zone

The vertical zone is referred to whether the damage is ground-level or atmospheric. Surface pollutants cause damage by concentrations of the pollutant accumulating near the Earth's surface Global pollutants cause damage by concentrations in the atmosphere

Sources and Causes

Air pollution comes from both natural and human-made (anthropogenic) sources. However, globally human-made pollutants from combustion, construction, mining, agriculture and warfare are increasingly significant in the air pollution equation. Motor vehicle emissions are one of the leading causes of air pollution. China, United States, Russia, India Mexico, and Japan are the world leaders in air pollution emissions.

Principal stationary pollution sources include chemical plants, coal-fired power plants, oil refineries, petrochemical plants, nuclear waste disposal activity, incinerators, large livestock farms (dairy cows, pigs, poultry, etc.), PVC factories, metals production factories, plastics factories, and other heavy industry. Agricultural air pollution comes from contemporary practices which include clear felling and burning of natural vegetation as well as spraying of pesticides and herbicides About 400 million metric tons of hazardous wastes are generated each year. The United States alone produces about 250 million metric tons.

Americans constitute less than 5% of the world's population, but produce roughly 25% of the world's CO_2, and generate approximately 30% of world's waste. In 2007, China has overtaken the United States as the world's biggest producer of CO_2, while still far behind based on per capita pollution - ranked 78th among the world's nations.

In February 2007, a report by the Intergovernmental Panel on Climate Change (IPCC), representing the work of 2,500 scientists, economists, and policymakers from more than 120 countries, said that humans have been the primary cause of global warming since 1950. Humans have ways to cut greenhouse gas emissions and avoid the consequences of global warming, a major climate report concluded. But to change the climate, the transition from fossil fuels like coal and oil needs to occur within decades, according to the final report this year from the UN's Intergovernmental Panel on Climate Change (IPCC).

Some of the more common soil contaminants are chlorinated hydrocarbons (CFH), heavy metals (such as chromium, cadmium–found in rechargeable batteries, and lead–found in lead paint, aviation fuel and still in some countries, gasoline), MTBE, zinc, arsenic and benzene. In 2001 a series of press reports culminating in a book called *Fateful Harvest* unveiled a widespread practice of recycling industrial byproducts into fertilizer, resulting in the contamination of the soil with various metals.

Ordinary municipal landfills are the source of many chemical substances entering the soil environment (and often groundwater), emanating from the wide variety of refuse accepted, especially substances illegally discarded there, or from pre-1970 landfills that may have been subject to little control in the U.S. or EU. There have also been some unusual releases of polychlorinated dibenzodioxins, commonly called *dioxins* for simplicity, such as TCDD.

Pollution can also be the consequence of a natural disaster. For example, hurricanes often involve water contamination from sewage, and petrochemical spills from ruptured boats or automobiles. Larger scale and environmental damage is not uncommon when coastal oil rigs or refineries are involved. Some sources of pollution, such as nuclear power plants or oil tankers, can produce widespread and potentially hazardous releases when accidents occur. In the case of noise pollution the dominant source class is the motor vehicle, producing about ninety percent of all unwanted noise worldwide.

Effects

Human Health: Adverse air quality can kill many organisms including humans. Ozone pollution can cause respiratory disease, cardiovascular disease, throat inflammation, chest pain, and congestion. Water pollution causes approximately 14,000 deaths per day, mostly due to contamination of drinking water by untreated sewage in developing countries.

An estimated 700 million Indians have no access to a proper toilet, and 1,000 Indian children die of diarrhea every day. Nearly 500 million Chinese lack access to safe drinking water. 656,000 people die prematurely each year in China because of air pollution. In India, air pollution is believed to cause 527,700 fatalities a year. Studies have estimated that the number of people killed annually in the US could be over 50,000.

Oil spills can cause skin irritations and rashes. Noise pollution induces hearing loss, high blood pressure, stress, and sleep disturbance. Mercury has been linked to developmental deficits in children and neurologic symptoms. Older people are majorly exposed to diseases induced by air pollution. Those with heart or lung disorders are under additional risk. Children and infants are also at serious risk. Lead and other heavy metals have been shown to cause neurological problems. Chemical and radioactive substances can cause cancer and as well as birth defects.

Environment

Pollution has been found to be present widely in the environment. There are a number of effects of this:

- Biomagnification describes situations where toxins (such as heavy metals) may pass through trophic levels, becoming exponentially more concentrated in the process.
- Carbon dioxide emissions cause ocean acidification, the ongoing decrease in the pH of the Earth's oceans as CO_2 becomes dissolved.
- The emission of greenhouse gases leads to global warming which affects ecosystems in many ways.
- Invasive species can out compete native species and reduce biodiversity. Invasive plants can contribute debris and biomolecules (allelopathy) that can alter soil and chemical

compositions of an environment, often reducing native species competitiveness.

- Nitrogen oxides are removed from the air by rain and fertilize land which can change the species composition of ecosystems.
- Smog and haze can reduce the amount of sunlight received by plants to carry out photosynthesis and leads to the production of tropospheric ozone which damages plants.
- Soil can become infertile and unsuitable for plants. This will affect other organisms in the food web.
- Sulphur dioxide and nitrogen oxides can cause acid rain which lowers the pH value of soil.

Environmental Health Information

The Toxicology and Environmental Health Information Programme (TEHIP) at the United States National Library of Medicine (NLM) maintains a comprehensive toxicology and environmental health web site that includes access to resources produced by TEHIP and by other government agencies and organizations. This web site includes links to databases, bibliographies, tutorials, and other scientific and consumer-oriented resources. TEHIP also is responsible for the Toxicology Data Network (TOXNET) an integrated system of toxicology and environmental health databases that are available free of charge on the web.

TOXMAP is a Geographic Information System (GIS) that is part of TOXNET. TOXMAP uses maps of the United States to help users visually explore data from the United States Environmental Protection Agency's (EPA) Toxics Release Inventory and Superfund Basic Research Programmes.

Regulation and Monitoring

To protect the environment from the adverse effects of pollution, many nations worldwide have enacted legislation to regulate various types of pollution as well as to mitigate the adverse effects of pollution.

Pollution Control

Pollution control is a term used in environmental management. It means the control of emissions and effluents into air, water or soil. Without pollution control, the waste products from consumption, heating, agriculture, mining, manufacturing, transportation and other human activities, whether they accumulate or disperse, will degrade the environment. In the hierarchy of controls, pollution prevention and waste minimization are more desirable than pollution control. In the field of land development, low impact development is a similar technique for the prevention of urban runoff.

Practices

- recycling
- reusing
- Waste minimisation
- mitigating
- preventing
- compost

Pollution Control Devices

- Dust collection systems
    - Baghouses
    - Cyclones
    - Electrostatic precipitators
- Scrubbers
    - Baffle spray scrubber
    - Cyclonic spray scrubber
    - Ejector venturi scrubber
    - Mechanically aided scrubber
    - Spray tower
    - Wet scrubber
- Sewage treatment
- Sedimentation (Primary treatment)
- Activated sludge biotreaters (Secondary treatment; also used for industrial wastewater)
- Aerated lagoons
- Constructed wetlands (also used for urban runoff)
- Industrial wastewater treatment
- API oil water separators
- Biofilters
- Dissolved air flotation (DAF)
- Powdered activated carbon treatment

- Ultrafiltration
- Vapor recovery systems
- Phytoremediation

Perspectives

The earliest precursor of pollution generated by life forms would have been a natural function of their existence. The attendant consequences on viability and population levels fell within the sphere of natural selection. These would have included the demise of a population locally or ultimately, species extinction. Processes that were untenable would have resulted in a new balance brought about by changes and adaptations. At the extremes, for any form of life, consideration of pollution is superseded by that of survival.

For humankind, the factor of technology is a distinguishing and critical consideration, both as an enabler and an additional source of byproducts. Short of survival, human concerns include the range from quality of life to health hazards. Since science holds experimental demonstration to be definitive, modern treatment of toxicity or environmental harm involves defining a level at which an effect is observable.

Common examples of fields where practical measurement is crucial include automobile emissions control, industrial exposure (e.g. Occupational Safety and Health Administration (OSHA) PELs), toxicology (e.g. LD_{50}), and medicine (e.g. medication and radiation doses). "The solution to pollution is dilution", is a dictum which summarizes a traditional approach to pollution management whereby sufficiently diluted pollution is not harmful. It is well-suited to some other modern, locally scoped applications such as laboratory safety procedure and hazardous material release emergency management. But it assumes that the dilutant is in virtually unlimited supply for the application or that resulting dilutions are acceptable in all cases.

Such simple treatment for environmental pollution on a wider scale might have had greater merit in earlier centuries when physical survival was often the highest imperative, human population and densities were lower, technologies were simpler and their byproducts more benign. But these are often no longer the case. Furthermore, advances have enabled measurement of concentrations not possible before. The use of statistical methods in evaluating outcomes has given currency to the principle of probable harm in cases where assessment is warranted but resorting to deterministic models is impractical or infeasible. In addition, consideration of the environment beyond direct impact on human beings has gained prominence.

Yet in the absence of a superseding principle, this older approach predominates practices throughout the world. It is the basis by which to gauge concentrations of effluent for legal release, exceeding which penalties are assessed or restrictions applied. One such superseding principle is contained in modern hazardous waste laws in developed countries, as the process of diluting hazardous waste to make it non-hazardous is usually a regulated treatment process. Migration from pollution dilution to elimination in many cases can be confronted by challenging economical and technological barriers.

Greenhouse Gases and Global Warming

Carbon dioxide, while vital for photosynthesis, is sometimes referred to as pollution, because raised levels of the gas in the atmosphere are affecting the Earth's climate. Disruption of the environment can also highlight the connection between areas of pollution that would normally be classified separately, such as those of water and air. Recent studies have investigated the potential for long-term rising levels of atmospheric carbon dioxide to cause slight but critical increases in the acidity of ocean waters, and the possible effects of this on marine ecosystems.

Most Polluted Places in the Developing World

The Blacksmith Institute, an international non-for-profit organization dedicated to eliminating life-threatening pollution in the developing world, issues an annual list of some of the world's worst polluted places. In the 2007 issues the ten top nominees, already industrialized countries excluded, are located in Azerbaijan, China, India, Peru, Russia, Ukraine and Zambia

Global Warming

Global warming is the rise in the average temperature of Earth's atmosphere and oceans since the late 19th century and its projected continuation. Since the early 20th century, Earth's mean surface temperature has increased by about 0.8 °C (1.4 °F), with about two-thirds of the increase occurring since 1980. Warming of the climate system is unequivocal, and scientists are more than 90% certain that it is primarily caused by increasing concentrations of greenhouse gases produced by human activities such as the burning of fossil fuels and deforestation. These findings are recognized by the national science academies of all major industrialized nations.

Climate model projections were summarized in the 2007 Fourth Assessment Report (AR4) by the Intergovernmental Panel on Climate Change (IPCC).

They indicated that during the 21st century the global surface temperature is likely to rise a further 1.1 to 2.9 °C (2 to 5.2 °F) for their lowest emissions scenario and 2.4 to 6.4 °C (4.3 to 11.5 °F) for their highest. The ranges of these estimates arise from the use of models with differing sensitivity to greenhouse gas concentrations.

Future warming and related changes will vary from region to region around the globe. The effects of an increase in global temperature include a rise in sea levels and a change in the amount and pattern of precipitation, as well a probable expansion of subtropical deserts. Warming is expected to be strongest in the Arctic and would be associated with the continuing retreat of glaciers, permafrost and sea ice.

Other likely effects of the warming include a more frequent occurrence of extreme-weather events including heat waves, droughts and heavy rainfall, ocean acidification and species extinctions due to shifting temperature regimes. Effects significant to humans include the threat to food security from decreasing crop yields and the loss of habitat from inundation.

Proposed policy responses to global warming include mitigation by emissions reduction, adaptation to its effects, and possible future geoengineering. Most countries are parties to the United Nations Framework Convention on Climate Change (UNFCCC), whose ultimate objective is to prevent dangerous anthropogenic (i.e., human-induced) climate change. Parties to the UNFCCC have adopted a range of policies designed to reduce greenhouse gas emissions and to assist in adaptation to global warming. Parties to the UNFCCC have agreed that deep cuts in emissions are required, and that future global warming should be limited to below 2.0 °C (3.6 °F) relative to the pre-industrial level. Reports published in 2011 by the United Nations Environment Programme and the International Energy Agency suggest that efforts as of the early 21st century to reduce emissions may be inadequate to meet the UNFCCC's 2 °C target.

Observed Temperature Changes

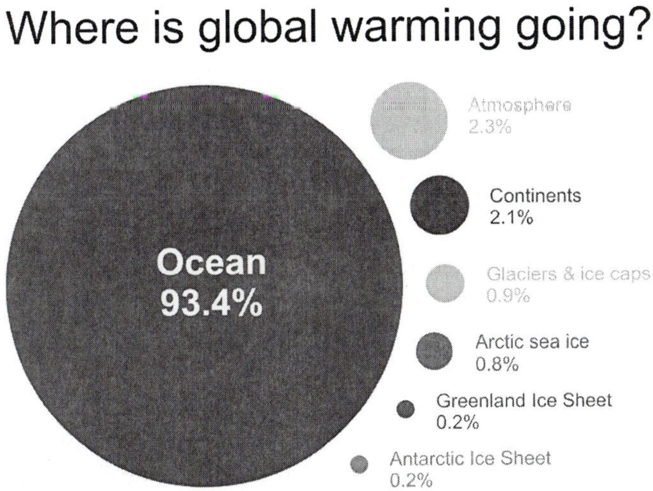

Figure: *The increase in ocean heat content is much larger than any other store of energy in the Earth's heat balance over the two periods 1961 to 2003 and 1993 to 2003, and accounts for more than 90% of the possible increase in heat content of the Earth system during these periods.*

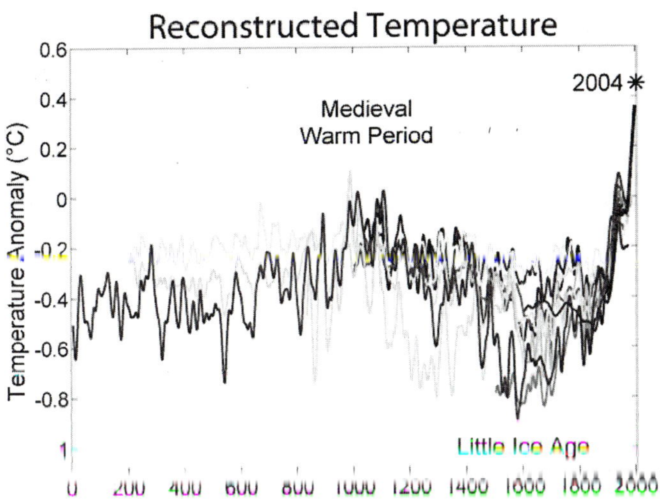

Figure: *Two millennia of mean surface temperatures according to different reconstructions from climate proxies, each smoothed on a decadal scale, with the instrumental temperature record overlaid in black.*

The Earth's average surface temperature rose by 0.74±0.18 °C over the period 1906–2005. The rate of warming over the last half of that period was almost double that for the period as a whole (0.13±0.03 °C per decade, versus 0.07±0.02 °C per decade). The urban heat island effect is very small, estimated to account for less than 0.002 °C of warming per decade since 1900. Temperatures in the lower troposphere have increased between 0.13 and 0.22 °C (0.22 and 0.4 °F) per decade since 1979, according to satellite temperature measurements. Climate proxies show the temperature to have been relatively stable over the one or two thousand years before 1850, with regionally varying fluctuations such as the Medieval Warm Period and the Little Ice Age.

The warming that is evident in the instrumental temperature record is consistent with a wide range of observations, as documented by many independent scientific groups. Examples include sea level rise (water expands as it warms), widespread melting of snow and ice, increased heat content of the oceans, increased humidity, and the earlier timing of spring events, e.g., the flowering of plants. The probability that these changes could have occurred by chance is virtually zero. Recent estimates by NASA's Goddard Institute for Space Studies (GISS) and the National Climatic Data Center show that 2005 and 2010 tied for the planet's warmest year since reliable, widespread instrumental measurements became available in the late 19th century, exceeding 1998 by a few hundredths of a degree. Estimates by the Climatic Research Unit (CRU) show 2005 as the second warmest year, behind 1998 with 2003 and 2010 tied for third warmest year, however, "the error estimate for individual years ... is at least ten times larger than the differences between these three years." The World Meteorological Organization (WMO) *statement on the status of the global climate in 2010* explains that, "The 2010 nominal value of +0.53 °C ranks just ahead of those of 2005 (+0.52 °C) and 1998 (+0.51 °C), although the differences between the three years are not statistically significant..."

Temperatures in 1998 were unusually warm because global temperatures are affected by the El Niño-Southern Oscillation (ENSO), and the strongest El Niño in the past century occurred during that year. Global temperature is subject to short-term fluctuations that overlay long term trends and can temporarily mask them.

The relative stability in temperature from 2002 to 2009 is consistent with such an episode. 2010 was also an El Niño year. On the low swing of the oscillation, 2011 as an La Niña year was cooler but it was still the 11th warmest year since records began in 1880. Of the 13 warmest years since 1880, 11 were the years from 2001 to

2011. Over the more recent record, 2011 was the warmest La Niña year in the period from 1950 to 2011, and was close to 1997 which was not at the lowest point of the cycle.

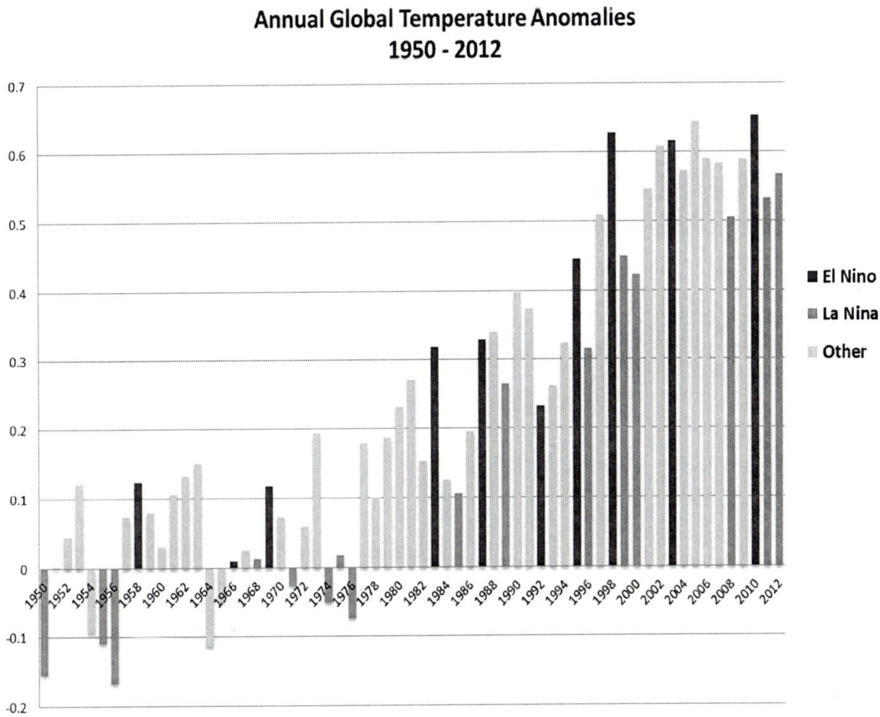

Figure: *NOAA graph of Global Annual Temperature Anomalies 1950–2012, showing the El Niño-Southern Oscillation*

Temperature changes vary over the globe. Since 1979, land temperatures have increased about twice as fast as ocean temperatures (0.25 °C per decade against 0.13 °C per decade).

Ocean temperatures increase more slowly than land temperatures because of the larger effective heat capacity of the oceans and because the ocean loses more heat by evaporation.

The northern hemisphere is also naturally warmer than the southern hemisphere mainly because of meridional heat transport in the oceans which has a differential of about 0.9 petawatts northwards, with an additional contribution from the albedo differences between the polar regions. Since the beginning of industrialisation the interhemispheric temperature difference has increased due to melting of sea ice and snow in the North.

Average arctic temperatures have been increasing at almost twice the rate of the rest of the world in the past 100 years, however arctic temperatures are also highly variable. Although more greenhouse gases are emitted in the Northern than Southern Hemisphere this does not contribute to the difference in warming because the major greenhouse gases persist long enough to mix between hemispheres.

The thermal inertia of the oceans and slow responses of other indirect effects mean that climate can take centuries or longer to adjust to changes in forcing. Climate commitment studies indicate that even if greenhouse gases were stabilized at 2000 levels, a further warming of about 0.5 °C (0.9 °F) would still occur.

Initial Causes of Temperature Changes (External Forcings)

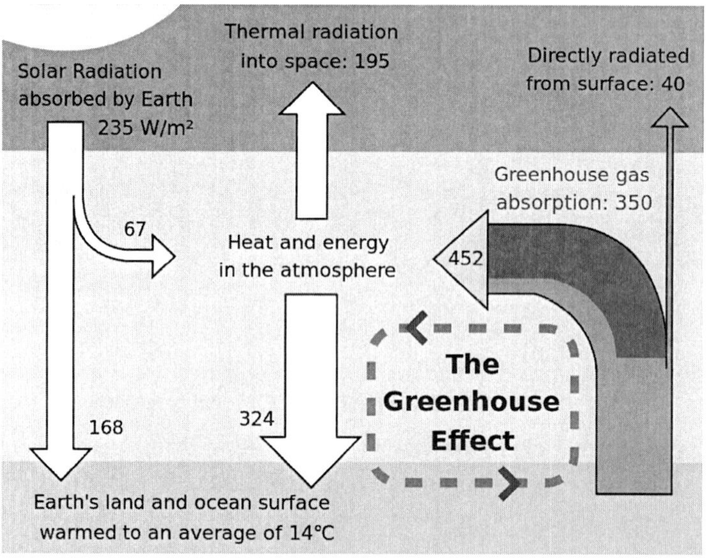

Figure: *Greenhouse effect schematic showing energy flows between space, the atmosphere, and earth's surface. Energy exchanges are expressed in watts per square meter (W/m²).*

The climate system can respond to changes in *external forcings*. External forcings can "push" the climate in the direction of warming or cooling. Examples of external forcings include changes in atmospheric composition (e.g., increased concentrations of greenhouse gases), solar luminosity, volcanic eruptions, and variations in Earth's orbit around the Sun. Orbital cycles vary slowly over tens of thousands of years and at present are in an overall cooling trend which would be expected to lead towards an ice age, but the 20th century instrumental temperature record shows a sudden rise in global temperatures.

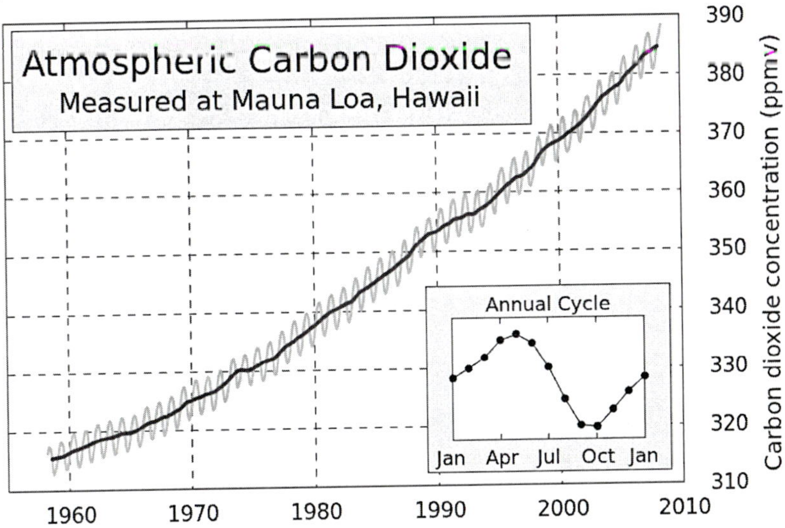

Figure: *This graph, known as the Keeling Curve, shows the increase of atmospheric carbon dioxide (CO_2) concentrations from 1958–2008. Monthly CO_2 measurements display seasonal oscillations in an upward trend; each year's maximum occurs during the Northern Hemisphere's late spring, and declines during its growing season as plants remove some atmospheric CO_2.*

Greenhouse Gases

The greenhouse effect is the process by which absorption and emission of infrared radiation by gases in the atmosphere warm a planet's lower atmosphere and surface. It was proposed by Joseph Fourier in 1824, discovered in 1860 by John Tyndall and was first investigated quantitatively by Svante Arrhenius in 1896.

Bubble diagram showing the share of global cumulative energy-related carbon dioxide emissions for major emitters between 1890-2007.

Naturally occurring amounts of greenhouse gases have a mean warming effect of about 33 °C (59 °F). The major greenhouse gases are water vapor, which causes about 36–70% of the greenhouse effect; carbon dioxide (CO_2), which causes 9–26%; methane (CH_4), which causes 4–9%; and ozone (O_3), which causes 3–7%. Clouds also affect the radiation balance through cloud forcings similar to greenhouse gases. Human activity since the Industrial Revolution has increased the amount of greenhouse gases in the atmosphere, leading to increased radiative forcing from CO_2, methane, tropospheric ozone, CFCs and nitrous oxide. The concentrations of CO_2 and methane have increased by 36% and 148% respectively since 1750. These levels are much higher than at any time during the last 800,000 years, the period for which reliable data has been extracted from ice cores. Less direct geological evidence indicates that CO_2 values higher than this were last seen about 20 million years ago.

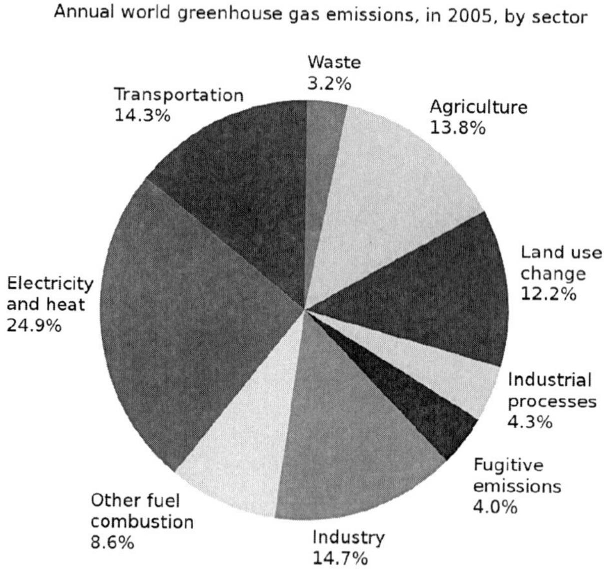

Figure: *Annual world greenhouse gas emissions, in 2005, by sector.*

Fossil fuel burning has produced about three-quarters of the increase in CO_2 from human activity over the past 20 years. The rest of this increase is caused mostly by changes in land-use, particularly deforestation. Estimates of global CO_2 emissions in 2011 from fossil fuel combustion, including cement production and gas flaring, was 34.8 billion tonnes (9.5 ± 0.5 PgC), an increase of 54% above emissions in 1990. Coal burning was responsible for 43% of the total emissions, oil 34%, gas 18%, cement 4.9% and gas flaring 0.7% Over the last three decades of the 20th century, gross domestic product per capita and population growth were the main drivers of increases in greenhouse gas emissions. CO_2 emissions are continuing to rise due to the burning of fossil fuels and land-use change. Emissions can be attributed to different regions. Attribution of emissions due to land-use change is a controversial issue.

Emissions scenarios, estimates of changes in future emission levels of greenhouse gases, have been projected that depend upon uncertain economic, sociological, technological, and natural developments. In most scenarios, emissions continue to rise over the century, while in a few, emissions are reduced. Fossil fuel reserves are abundant, and will not limit carbon emissions in the 21st century. Emission scenarios, combined with modelling of the carbon cycle, have been used to produce estimates of how atmospheric concentrations of greenhouse gases might change in the future.

Using the six IPCC SRES "marker" scenarios, models suggest that by the year 2100, the atmospheric concentration of CO_2 could range between 541 and 970 ppm. This is an increase of 90–250% above the concentration in the year 1750. The popular media and the public often confuse global warming with ozone depletion, i.e., the destruction of stratospheric ozone by chlorofluorocarbons. Although there are a few areas of linkage, the relationship between the two is not strong. Reduced stratospheric ozone has had a slight cooling influence on surface temperatures, while increased tropospheric ozone has had a somewhat larger warming effect.

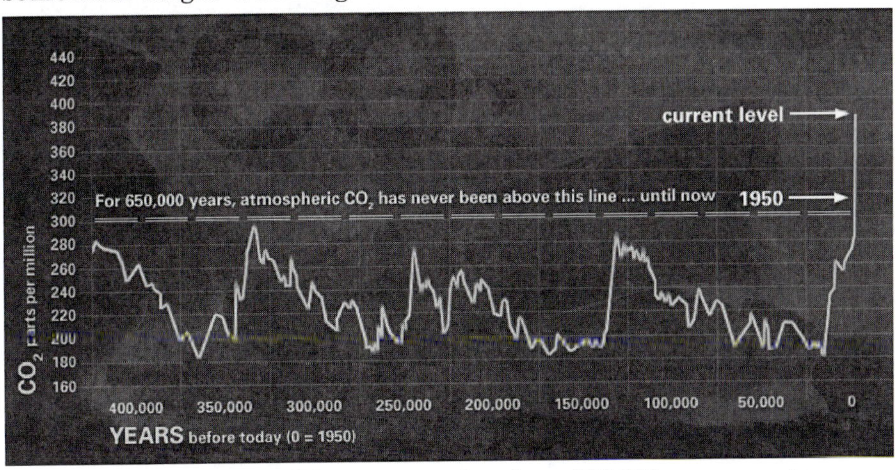

Figure: Atmospheric CO_2 concentration from 650,000 years ago to near present, using ice core proxy data and direct measurements

Particulates and Soot

Global dimming, a gradual reduction in the amount of global direct irradiance at the Earth's surface, was observed from 1961 until at least 1990. The main cause of this dimming is particulates produced by volcanoes and human made pollutants, which exerts a cooling effect by increasing the reflection of incoming sunlight. The effects of the products of fossil fuel combustion – CO_2 and aerosols – have largely offset one another in recent decades, so that net warming has been due to the increase in non-CO_2 greenhouse gases such as methane. Radiative forcing due to particulates is temporally limited due to wet deposition which causes them to have an atmospheric lifetime of one week. Carbon dioxide has a lifetime of a century or more, and as such, changes in particulate concentrations will only delay climate changes due to carbon dioxide.

In addition to their direct effect by scattering and absorbing solar radiation, particulates have indirect effects on the radiation budget. Sulphates act as cloud condensation nuclei and thus lead to clouds that have more and smaller cloud droplets. These clouds reflect solar radiation more efficiently than clouds with fewer and larger droplets, known as the Twomey effect. This effect also causes droplets to be of more uniform size, which reduces growth of raindrops and makes the cloud more reflective to incoming sunlight, known as the Albrecht effect. Indirect effects are most noticeable in marine stratiform clouds, and have very little radiative effect on convective clouds. Indirect effects of particulates represent the largest uncertainty in radiative forcing. Soot may cool or warm the surface, depending on whether it is airborne or deposited. Atmospheric soot directly absorb solar radiation, which heats the atmosphere and cools the surface. In isolated areas with high soot production, such as rural India, as much as 50% of surface warming due to greenhouse gases may be masked by atmospheric brown clouds.

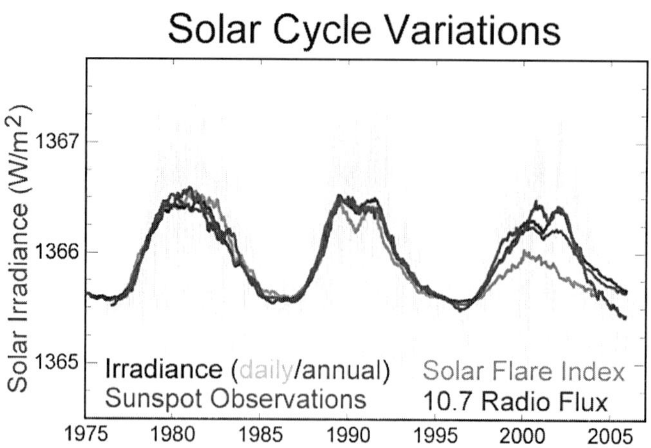

***Figure:** Satellite observations of Total Solar Irradiance from 1979–2006.*

When deposited, especially on glaciers or on ice in arctic regions, the lower surface albedo can also directly heat the surface. The influences of particulates, including black carbon, are most pronounced in the tropics and sub-tropics, particularly in Asia, while the effects of greenhouse gases are dominant in the extratropics and southern hemisphere.

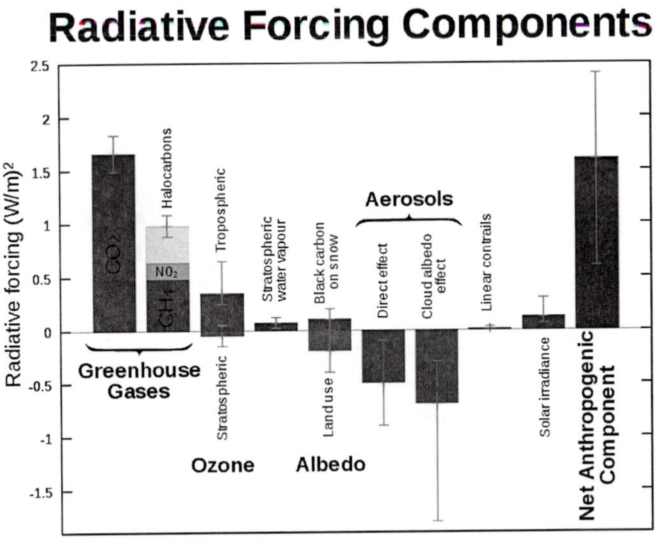

Figure: *Contribution of natural factors and human activities to radiative forcing of climate change. Radiative forcing values are for the year 2005, relative to the pre-industrial era (1750). The contribution of solar irradiance to radiative forcing is 5% the value of the combined radiative forcing due to increases in the atmospheric concentrations of carbon dioxide, methane and nitrous oxide.*

Solar Activity

Since 1978, output from the Sun has been precisely measured by satellites. These measurements indicate that the Sun's output has not increased since 1978, so the warming during the past 30 years cannot be attributed to an increase in solar energy reaching the Earth. In the three decades since 1978, the combination of solar and volcanic activity probably had a slight cooling influence on the climate. Climate models have been used to examine the role of the sun in recent climate change. Models are unable to reproduce the rapid warming observed in recent decades when they only take into account variations in solar output and volcanic activity. Models are, however, able to simulate the observed 20th century changes in temperature when they include all of the most important external forcings, including human influences and natural forcings.

Another line of evidence against the sun having caused recent climate change comes from looking at how temperatures at different levels in the Earth's atmosphere have changed. Models and observations show that greenhouse warming results in warming of the lower atmosphere (called the troposphere) but cooling of the upper atmosphere (called the stratosphere). Depletion of the ozone layer by chemical refrigerants has also resulted in a strong cooling effect in the stratosphere. If the sun was responsible for observed warming, warming of both the troposphere and stratosphere would be expected.

Feedback

The climate system includes a range of *feedbacks* which alter the response of the system to changes in external forcings. Positive feedbacks increase the response of the climate system to an initial forcing, while negative feedbacks reduce the response of the climate system to an initial forcing. There are a range of feedbacks in the climate system, including water vapor, changes in ice-albedo (snow and ice cover affect how much the Earth's surface absorbs or reflects incoming sunlight), clouds, and changes in the Earth's carbon cycle (e.g., the release of carbon from soil). The main negative feedback is the energy which the Earth's surface radiates into space as infrared radiation. According to the Stefan-Boltzmann law, if temperature doubles, radiated energy increases by a factor of 16 (2 to the 4th power). Feedbacks are an important factor in determining the sensitivity of the climate system to increased atmospheric greenhouse gas concentrations. Other factors being equal, a higher *climate sensitivity* means that more warming will occur for a given increase in greenhouse gas forcing. Uncertainty over the effect of feedbacks is a major reason why different climate models project different magnitudes of warming for a given forcing scenario. More research is needed to understand the role of clouds and carbon cycle feedbacks in climate projections. The IPCC projections given in the lede span the "likely" range (greater than 66% probability, based on expert judgement) for the selected emissions scenarios. However, the IPCC's projections do not reflect the full range of uncertainty. The lower end of the "likely" range appears to be better constrained than the upper end of the "likely" range.

Climate Models

Projected change in annual mean surface air temperature from the late 20th century to the middle 21st century, based on a medium emissions scenario (SRES A1B). This scenario assumes that no future policies are adopted to limit greenhouse gas emissions. Image credit: NOAA GFDL. A climate model is a computerized representation of the five components of the climate system: Atmosphere, hydrosphere, cryosphere, land surface, and biosphere. Such models are based on physical principles including fluid dynamics, thermodynamics and radiative transfer. There can be components which represent air movement, temperature, clouds, and other atmospheric properties; ocean temperature, salt content, and circulation; ice cover on land and sea; the transfer of heat and moisture from soil and vegetation to the atmosphere; chemical and biological processes; and others.

Although researchers attempt to include as many processes as possible, simplifications of the actual climate system are inevitable because of the constraints of available computer power and limitations in knowledge of the climate system. Results from models can also vary due to different greenhouse gas inputs and the model's climate sensitivity. For example, the uncertainty in IPCC's 2007 projections is caused by (1) the use of multiple models with differing sensitivity to greenhouse gas concentrations, (2) the use of differing estimates of humanities' future greenhouse gas emissions, (3) any additional emissions from climate feedbacks that were not included in the models IPCC used to prepare its report, i.e., greenhouse gas releases from permafrost. The models do not assume the climate will warm due to increasing levels of greenhouse gases. Instead the models predict how greenhouse gases will interact with radiative transfer and other physical processes. One of the mathematical results of these complex equations is a prediction whether warming or cooling will occur. Recent research has called special attention to the need to refine models with respect to the effect of clouds and the carbon cycle.

Models are also used to help investigate the causes of recent climate change by comparing the observed changes to those that the models project from various natural and human-derived causes. Although these models do not unambiguously attribute the warming that occurred from approximately 1910 to 1945 to either natural variation or human effects, they do indicate that the warming since 1970 is dominated by man-made greenhouse gas emissions. The physical realism of models is tested by examining their ability to simulate contemporary or past climates. Climate models produce a good match to observations of global temperature changes over the last century, but do not simulate all aspects of climate. Not all effects of global warming are accurately predicted by the climate models used by the IPCC. Observed Arctic shrinkage has been faster than that predicted. Precipitation increased proportional to atmospheric humidity, and hence significantly faster than global climate models predict.

Observed and Expected Environmental Effects

"Detection" is the process of demonstrating that climate has changed in some defined statistical sense, without providing a reason for that change. Detection does not imply attribution of the detected change to a particular cause. "Attribution" of causes of climate change is the process of establishing the most likely causes for the detected change with some defined level of confidence. Detection and attribution may also be applied to observed changes in physical, ecological and social systems.

Natural Systems

Global warming has been detected in a number of natural systems. Some of these changes are described in the section on observed temperature changes, e.g., sea level rise and widespread decreases in snow and ice extent. Anthropogenic forcing has likely contributed to some of the observed changes, including sea level rise, changes in climate extremes (such as the number of warm and cold days), declines in Arctic sea ice extent, and to glacier retreat. Projections of sea level rise over the 21st century are shown opposite. On the timescale of centuries to millennia, the melting of ice sheets could result in even higher sea level rise. Partial deglaciation of the Greenland ice sheet, and possibly the West Antarctic Ice Sheet, could contribute 4–6 metres (13 to 20 ft) or more to sea level rise. Changes in regional climate are expected to include greater warming over land, with most warming at high northern latitudes, and least warming over the Southern Ocean and parts of the North Atlantic Ocean. During the 21st century, glaciers and snow cover are projected to continue their widespread retreat. Projections of declines in Arctic sea ice vary. Recent projections suggest that Arctic summers could be ice-free (defined as ice extent less than 1 million square km) as early as 2025-2030.

Future changes in precipitation are expected to follow existing trends, with reduced precipitation over subtropical land areas, and increased precipitation at subpolar latitudes and some equatorial regions. Projections suggest a probable increase in the frequency and severity of some extreme weather events, such as heat waves.

Ecological Systems

In terrestrial ecosystems, the earlier timing of spring events, and poleward and upward shifts in plant and animal ranges, have been linked with high confidence to recent warming. Future climate change is expected to particularly affect certain ecosystems, including tundra, mangroves, and coral reefs. It is expected that most ecosystems will be affected by higher atmospheric CO_2 levels, combined with higher global temperatures. Overall, it is expected that climate change will result in the extinction of many species and reduced diversity of ecosystems. Dissolved CO_2 increases ocean acidity. This process is known as ocean acidification and has been called the "equally evil twin" of global climate change. Increased ocean acidity decreases the amount of carbonate ions, which organisms at the base of the marine food chain, such as foraminifera, use to make structures they need to survive. The current rate of ocean acidification is many times faster than at least the past 300 million years, which included four mass extinctions that involved rising ocean acidity, such as the Permian mass extinction, which killed 95% of marine species. By the end of the century, acidity changes since the industrial revolution would match the Palaeocene-Eocene Thermal Maximum, which occurred over 5000 years and killed 35–50% of benthic foraminifera.

Large-scale and Abrupt Impacts

Climate change could result in global, large-scale changes in natural and social systems. Two examples are ocean acidification caused by increased atmospheric concentrations of carbon dioxide, and the long-term melting of ice sheets, which contributes to sea level rise. Some large-scale changes could occur abruptly, i.e., over a short time period, and might also be irreversible. An example of abrupt climate change is the rapid release of methane from permafrost, which would lead to amplified global warming. Scientific understanding of abrupt climate change is generally poor. However, the probability of abrupt changes appears to be very low. Factors that may increase the probability of abrupt climate change include higher magnitudes of global warming, warming that occurs more rapidly, and warming that is sustained over longer time periods.

Observed and Expected Effects on Social Systems

Vulnerability of human societies to climate change mainly lies in the effects of extreme-weather events rather than gradual climate change. Impacts of climate change so far include adverse effects on small islands, adverse effects on indigenous populations in high-latitude areas, and small but discernable effects on human health Over the 21st century, climate change is likely to adversely affect hundreds of millions of people through increased coastal flooding, reductions in water supplies, increased malnutrition and increased health impacts. Most economic studies suggest losses of world gross domestic product (GDP) for this magnitude of warming.

Food Security

Under present trends, by 2030, maize production in Southern Africa could decrease by up to 30% while rice, millet and maize in South Asia could decrease by up to 10%. By 2080, yields in developing countries could decrease by 10% to 25% on average while India could see a drop of 30% to 40%. By 2100, while the population of three billion is expected to double, rice and maize yields in the tropics are expected to decrease by 20–40% because of higher temperatures without accounting for the decrease in yields as a result of soil moisture and water supplies stressed by rising temperatures.

Future warming of around 3 °C (by 2100, relative to 1990–2000) could result in increased crop yields in mid- and high-latitude areas, but in low-latitude areas, yields could decline, increasing the risk of malnutrition. A similar regional pattern of net benefits and costs could occur for economic (market-sector) effects. Warming above 3 °C could result in crop yields falling in temperate regions, leading to a reduction in global food production.

Habitat Inundation

In small islands and megadeltas, inundation as a result of sea level rise is expected to threaten vital infrastructure and human settlements. This could lead to issues of homelessness in countries with low lying areas such as Bangladesh, as well as statelessness for populations in countries such as the Maldives and Tuvalu.

Responses to Global Warming

Mitigation: Reducing the amount of future climate change is called mitigation of climate change. The IPCC defines mitigation as activities that reduce greenhouse gas (GHG) emissions, or enhance the capacity of carbon sinks to absorb GHGs from the atmosphere. Studies indicate substantial potential for future reductions in emissions by a combination of emission-reducing activities such as energy conservation, increased energy efficiency, and satisfying more of society's power demands with renewable energy and/or nuclear energy sources. Climate mitigation also includes acts to enhance natural sinks, such as reforestation. In order to limit warming to within the lower range described in the IPCC's "Summary Report for Policymakers" it will be necessary to adopt policies that will limit greenhouse gas emissions to one of several significantly different scenarios described in the full report. This will become more and more difficult with each year of increasing volumes of emissions and even more drastic measures will be required in later years to stabilize a desired atmospheric concentration of greenhouse gases. Energy-related carbon-dioxide (CO_2) emissions in 2010 were the highest in history, breaking the prior record set in 2008.

Adaptation

Other policy responses include adaptation to climate change. Adaptation to climate change may be planned, either in reaction to or anticipation of climate change, or spontaneous, i.e., without government intervention. Planned adaptation is already occurring on a limited basis. The barriers, limits, and costs of future adaptation are not fully understood. A concept related to adaptation is "adaptive capacity," which is the ability of a system (human, natural or managed) to adjust to climate change (including climate variability and extremes) to moderate potential damages, to take advantage of opportunities, or to cope with consequences. Unmitigated climate change (i.e., future climate change without efforts to limit greenhouse gas emissions) would, in the long term, be likely to exceed the capacity of natural, managed and human systems to adapt.

Views on Global Warming

There are different views over what the appropriate policy response to climate change should be. These competing views weigh the benefits of limiting emissions of greenhouse gases against the costs. In general, it seems likely that climate change will impose greater damages and risks in poorer regions.

Scientific Opinion on Climate Change

The scientific opinion on climate change is that the Earth's climate system is unequivocally warming, and it is more than 90% certain that humans are causing most of it through activities that increase concentrations of greenhouse gases in the atmosphere, such as deforestation and burning fossil fuels. In addition, it is likely that some potential further greenhouse gas warming has been offset by increased aerosols. This scientific consensus is expressed in synthesis reports, by scientific bodies of national or international standing, and by surveys of opinion among climate scientists. Individual scientists, universities, and laboratories contribute to the overall scientific opinion via their peer-reviewed publications, and the areas of collective agreement and relative certainty are summarised in these high level reports and surveys. National and international science academies and scientific societies have assessed the current scientific opinion, in particular on recent global warming. These assessments have largely followed or endorsed the Intergovernmental Panel on Climate Change (IPCC) position of January 2001 which states:

An increasing body of observations gives a collective picture of a warming world and other changes in the climate system... There is new and stronger evidence that most of the warming observed over the last 50 years is attributable to human activities.

The main conclusions of the IPCC Working Group I on global warming were the following:

1. The global average surface temperature has risen 0.6 ± 0.2 °C since the late 19th century, and 0.17 °C per decade in the last 30 years.
2. "There is new and stronger evidence that most of the warming observed over the last 50 years is attributable to human activities", in particular emissions of the greenhouse gases carbon dioxide and methane.
3. If greenhouse gas emissions continue the warming will also continue, with temperatures projected to increase by 1.4 °C to 5.8 °C between 1990 and 2100. Accompanying this temperature increase will be increases in some types of extreme weather and a projected sea level rise. From IPCC Working Group II: On balance the impacts of global warming will be significantly negative, especially for larger values of warming.

No scientific body of national or international standing maintains a formal opinion dissenting from any of these three main points; the last was the American Association of Petroleum Geologists, which in 2007 updated its 1999 statement rejecting the likelihood of human influence on recent climate with its current non-committal position. Some other organizations, primarily those focusing on geology, also hold non-committal positions.

Synthesis Reports

Synthesis reports are assessments of scientific literature that compile the results of a range of stand-alone studies in order to achieve a broad level of understanding, or to describe the state of knowledge of a given subject.

Intergovernmental Panel on Climate Change (IPCC) 2007

In February 2007, the IPCC released a summary of the forthcoming Fourth Assessment Report. According to this summary, the Fourth Assessment Report finds that human actions are "very likely" the cause of global warming, meaning a 90% or greater probability. Global warming in this case is indicated by an increase of 0.75 degrees in average global temperatures over the last 100 years. *The New York Times* reported that "the leading international network of climate scientists has concluded for the first time that global warming is 'unequivocal' and that human activity is the main driver, very likely' causing most of the rise in temperatures since 1950".

A retired journalist for *The New York Times*, William K. Stevens wrote: "The Intergovernmental Panel on Climate Change said the likelihood was 90 percent to 99 percent that emissions of heat-trapping greenhouse gases like carbon dioxide, spewed from tailpipes and smokestacks, were the dominant cause of the observed warming of the last 50 years. In the panel's parlance, this level of certainty is labelled 'very likely'. Only rarely does scientific odds-making provide a more definite answer than that, at least in this branch of science, and it describes the endpoint, so far, of a progression.".

The Associated Press Summarized the Position on Sea Level Rise: On sea levels, the report projects rises of 7 to 23 inches by the end of the century. An additional 3.9 to 7.8 inches are possible if recent, surprising melting of polar ice sheets continues.

U.S. Global Change Research Programme

***Formerly the Climate Change Science Programme*:** The U.S. Global Change Research Programme reported in June, 2009 that: Observations show that warming of the climate is unequivocal. The global warming observed over the past 50 years is due primarily to human-induced emissions of heat-trapping gases. These emissions come mainly from the burning of fossil fuels (coal, oil, and gas), with important contributions from the clearing of forests, agricultural practices, and other activities.

The report, which is about the effects that climate change is having in the United States, also says:

Climate-related changes have already been observed globally and in the United States. These include increases in air and water temperatures, reduced frost days, increased frequency and intensity of heavy downpours, a rise in sea level, and reduced snow cover, glaciers, permafrost, and sea ice. A longer ice-free period on lakes and rivers, lengthening of the growing season, and increased water vapor in the atmosphere have also been observed. Over the past 30 years, temperatures have risen faster in winter than in any other season, with average winter temperatures in the Midwest and northern Great Plains increasing more than 7°F. Some of the changes have been faster than previous assessments had suggested.

Arctic Climate Impact Assessment

In 2004, the intergovernmental Arctic Council and the non-governmental International Arctic Science Committee released the synthesis report of the Arctic Climate Impact Assessment: Climate conditions in the past provide evidence that rising atmospheric carbon dioxide levels are associated with rising global temperatures. Human activities, primarily the burning of fossil fuels (coal, oil, and natural gas), and secondarily the clearing of land, have increased the concentration of carbon dioxide, methane, and other heat-trapping ("greenhouse") gases in the atmosphere...There is international scientific consensus that most of the warming observed over the last 50 years is attributable to human activities.

Policy

There is an extensive literature on what policies might be effective in responding to climate change. Some scientific bodies have recommended specific policies to governments (refer to the later sections of the article). The natural and social sciences can play a role in informing an effective response to climate change. However, policy decisions may require value judgements. For example, the US National Research Council has commented:

The question of whether there exists a "safe" level of concentration of greenhouse gases cannot be answered directly because it would require a value judgement of what constitutes an acceptable risk to human welfare and ecosystems in various parts of the world, as well as a more quantitative assessment of the risks and costs associated with the various impacts of global warming. In general, however, risk increases with increases in both the rate and the magnitude of climate change. This article mostly focuses on the views of natural scientists. However, social scientists, medical experts, engineers and philosophers have also commented on climate change science and policies. Climate change policy is discussed in several articles: climate change mitigation, climate change adaptation, geoengineering, politics of global warming, climate ethics, and economics of global warming.

Statements by Scientific Organizations of National or International Standing

This is a list of scientific bodies of national or international standing, that have issued formal statements of opinion, classifies those organizations according to whether they concur with the IPCC view, are non-committal, or dissent from it.

Concurring

Academies of Science (General Science) : A total of 34 national science academies have made formal declarations since 2001 confirming anthropogenic global warming and urging the nations of the world to reduce emissions of greenhouse gases. This includes 33 who have signed joint science academies' statements and one individual declaration made by the Polish Academy of Sciences made in 2007.

Joint National Science Academies' Statements

- 2001 Following the publication of the IPCC Third Assessment Report, seventeen national science academies issued a joint statement, entitled "The Science of Climate Change", explicitly acknowledging the IPCC position as representing the scientific consensus on climate change science. The statement, printed in an editorial in the journal *Science* on May 18, 2001, was signed by the science academies of Australia, Belgium, Brazil, Canada, the Caribbean, China, France, Germany, India, Indonesia, Ireland, Italy, Malaysia, New Zealand, Sweden, Turkey, and the United Kingdom.
- 2005 The national science academies of the G8 nations, plus Brazil, China and India, three of the largest emitters of greenhouse gases in the developing world, signed a statement on the global response to climate change. The statement stresses that the scientific understanding of climate change is now sufficiently clear to justify nations taking prompt action, and explicitly endorsed the IPCC consensus. The eleven signatories were the science academies of Brazil, Canada, China, France, Germany, India, Italy, Japan, Russia, the United Kingdom, and the United States.
- 2007 In preparation for the 33rd G8 summit, the national science academies of the G8+5 nations issued a declaration referencing the position of the 2005 joint science academies' statement, and acknowledging the confirmation of their previous conclusion by recent research. Following the IPCC

Fourth Assessment Report, the declaration states, "It is unequivocal that the climate is changing, and it is very likely that this is predominantly caused by the increasing human interference with the atmosphere. These changes will transform the environmental conditions on Earth unless counter-measures are taken." The thirteen signatories were the national science academies of Brazil, Canada, China, France, Germany, Italy, India, Japan, Mexico, Russia, South Africa, the United Kingdom, and the United States.

- 2007 In preparation for the 33rd G8 summit, the Network of African Science Academies submitted a joint "statement on sustainability, energy efficiency, and climate change" :

A consensus, based on current evidence, now exists within the global scientific community that human activities are the main source of climate change and that the burning of fossil fuels is largely responsible for driving this change. The IPCC should be congratulated for the contribution it has made to public understanding of the nexus that exists between energy, climate and sustainability. The thirteen signatories were the science academies of Cameroon, Ghana, Kenya, Madagascar, Nigeria, Senegal, South Africa, Sudan, Tanzania, Uganda, Zambia, Zimbabwe, as well as the African Academy of Sciences.

- 2008 In preparation for the 34th G8 summit, the national science academies of the G8+5 nations issued a declaration reiterating the position of the 2005 joint science academies' statement, and reaffirming "that climate change is happening and that anthropogenic warming is influencing many physical and biological systems." Among other actions, the declaration urges all nations to "(t)ake appropriate economic and policy measures to accelerate transition to a low carbon society and to encourage and effect changes in individual and national behaviour." The thirteen signatories were the same national science academies that issued the 2007 joint statement.
- 2009 In advance of the UNFCCC negotiations to be held in Copenhagen in December 2009, the national science academies of the G8+5 nations issued a joint statement declaring, "Climate change and sustainable energy supply are crucial challenges for the future of humanity. It is essential that world leaders agree on the emission reductions needed to combat negative consequences of anthropogenic climate change". The statement

references the IPCC's Fourth Assessment of 2007, and asserts that "climate change is happening even faster than previously estimated; global CO_2 emissions since 2000 have been higher than even the highest predictions, Arctic sea ice has been melting at rates much faster than predicted, and the rise in the sea level has become more rapid." The thirteen signatories were the same national science academies that issued the 2007 and 2008 joint statements.

Polish Academy of Sciences

In December 2007, the General Assembly of the Polish Academy of Sciences (PAN), which has not been a signatory to joint national science academies' statements issued a declaration endorsing the IPCC conclusions, and stating:

It is the duty of Polish science and the national government to, in a thoughtful, organized and active manner, become involved in realisation of these ideas. Problems of global warming, climate change, and their various negative impacts on human life and on the functioning of entire societies are one of the most dramatic challenges of modern times. PAS General Assembly calls on the national scientific communities and the national government to actively support Polish participation in this important endeavour.

Additional National Science Academy and Society Statements

- American Association for the Advancement of Science as the world's largest general scientific society, adopted an official statement on climate change in 2006:

The scientific evidence is clear: global climate change caused by human activities is occurring now, and it is a growing threat to society....The pace of change and the evidence of harm have increased markedly over the last five years. The time to control greenhouse gas emissions is now.

- Federation of Australian Scientific and Technological Societies in 2008 published *FASTS Statement on Climate Change* which states:

Global climate change is real and measurable...To reduce the global net economic, environmental and social losses in the face of these impacts, the policy objective must remain squarely focused on returning greenhouse gas concentrations to near pre-industrial levels through the reduction of emissions. The spatial and temporal fingerprint of warming can be traced to increasing greenhouse gas concentrations in the atmosphere, which are a direct result of burning fossil fuels, broad-scale deforestation and other human activity.

- United States National Research Council through its Committee on the Science of Climate Change in 2001, published *Climate Change Science: An Analysis of Some Key Questions*. This report explicitly endorses the IPCC view of attribution of recent climate change as representing the view of the scientific community:

The changes observed over the last several decades are likely mostly due to human activities, but we cannot rule out that some significant part of these changes is also a reflection of natural variability. Human-induced warming and associated sea level rises are expected to continue through the 21st century... The IPCC's conclusion that most of the observed warming of the last 50 years is likely to have been due to the increase in greenhouse gas concentrations accurately reflects the current thinking of the scientific community on this issue.

- Royal Society of New Zealand having signed onto the first joint science academies' statement in 2001, released a separate statement in 2008 in order to clear up "the controversy over climate change and its causes, and possible confusion among the public":

The globe is warming because of increasing greenhouse gas emissions. Measurements show that greenhouse gas concentrations in the atmosphere are well above levels seen for many thousands of years. Further global climate changes are predicted, with impacts expected to become more costly as time progresses. Reducing future impacts of climate change will require substantial reductions of greenhouse gas emissions.

- The Royal Society of the United Kingdom has not changed its concurring stance reflected in its participation in joint national science academies' statements on anthropogenic global warming. According to the Telegraph, "The most prestigious group of scientists in the country was forced to act after fellows complained that doubts over man made global warming were not being communicated to the public". In May 2010, it announced that it "is presently drafting a new public facing document on climate change, to provide an updated status report on the science in an easily accessible form, also addressing the levels of certainty of key components." The society says that it is three years since the last such document was published and that, after an extensive process of debate and review, the new document was printed in September 2010.

It summarises the current scientific evidence and highlights the areas where the science is well established, where there is still some debate, and where substantial uncertainties remain. The society has stated that "this is not the same as saying that the climate science itself is in error – no Fellows have expressed such a view to the RS". The introduction includes this statement:

There is strong evidence that the warming of the Earth over the last half-century has been caused largely by human activity, such as the burning of fossil fuels and changes in land use, including agriculture and deforestation.

International Science Academies

- African Academy of Sciences in 2007 was a signatory to the "statement on sustainability, energy efficiency, and climate change", the joint statement of African science academies, organized through the Network of African Science Academies, confirming anthropogenic global warming and presented to the leaders meeting at the G8 Summit in Heiligendamm, Germany.
- European Academy of Sciences and Arts in 2007 issued a formal declaration on climate change titled *Let's Be Honest*:

Human activity is most *likely* responsible for climate warming. Most of the climatic warming over the last 50 years is *likely* to have been caused by increased concentrations of greenhouse gases in the atmosphere. Documented long-term climate changes include changes in Arctic temperatures and ice, widespread changes in precipitation amounts, ocean salinity, wind patterns and extreme weather including droughts, heavy precipitation, heat waves and the intensity of tropical cyclones. The above development potentially has dramatic consequences for mankind's future.

- European Science Foundation in a 2007 position paper states:

There is now convincing evidence that since the industrial revolution, human activities, resulting in increasing concentrations of greenhouse gases have become a major agent of climate change... On-going and increased efforts to mitigate climate change through reduction in greenhouse gases are therefore crucial.

- InterAcademy Council As the representative of the world's scientific and engineering academies, the InterAcademy Council issued a report in 2007 titled *Lighting the Way: Toward a Sustainable Energy Future.*

Current patterns of energy resources and energy usage are proving detrimental to the long-term welfare of humanity. The integrity of essential natural systems is already at risk from climate change caused by the atmospheric emissions of greenhouse gases. Concerted efforts should be mounted for improving energy efficiency and reducing the carbon intensity of the world economy.

- International Council of Academies of Engineering and Technological Sciences (CAETS) in 2007, issued a *Statement on Environment and Sustainable Growth*:

As reported by the Intergovernmental Panel on Climate Change (IPCC), most of the observed global warming since the mid-20th century is very likely due to human-produced emission of greenhouse gases and this warming will continue unabated if present anthropogenic emissions continue or, worse, expand without control. CAETS, therefore, endorses the many recent calls to decrease and control greenhouse gas emissions to an acceptable level as quickly as possible.

Physical and Chemical Sciences

- American Chemical Society
- American Institute of Physics
- American Physical Society
- Australian Institute of Physics
- European Physical Society

Earth sciences

American Geophysical Union : The American Geophysical Union (AGU) statement, adopted by the society in 2003 and revised in 2007, affirms that rising levels of greenhouse gases have caused and will continue to cause the global surface temperature to be warmer:

The Earth's climate is now clearly out of balance and is warming. Many components of the climate system—including the temperatures of the atmosphere, land and ocean, the extent of sea ice and mountain glaciers, the sea level, the distribution of precipitation, and the length of seasons—are now changing at rates and in patterns that are not natural and are best explained by the increased atmospheric abundances of greenhouse gases and aerosols generated by human activity during the 20th century. Global average surface temperatures increased on average by about 0.6°C over the period 1956–2006. As of 2006, eleven of the previous twelve years were warmer than any others since 1850.

The observed rapid retreat of Arctic sea ice is expected to continue and lead to the disappearance of summertime ice within this century. Evidence from most oceans and all continents except Antarctica shows warming attributable to human activities. Recent changes in many physical and biological systems are linked with this regional climate change. A sustained research effort, involving many AGU members and summarized in the 2007 assessments of the Intergovernmental Panel on Climate Change, continues to improve our scientific understanding of the climate.

American Society of Agronomy, Crop Science Society of America, and Soil Science Society of America In May, 2011, the American Society of Agronomy (ASA), Crop Science Society of America (CSSA), and Soil Science Society of America (SSSA) issued a joint position statement on climate change as it relates to agriculture:

A comprehensive body of scientific evidence indicates beyond reasonable doubt that global climate change is now occurring and that its manifestations threaten the stability of societies as well as natural and managed ecosystems. Increases in ambient temperatures and changes in related processes are directly linked to rising anthropogenic greenhouse gas (GHG) concentrations in the atmosphere. Unless the emissions of GHGs are curbed significantly, their concentrations will continue to rise, leading to changes in temperature, precipitation, and other climate variables that will undoubtedly affect agriculture around the world. Climate change has the potential to increase weather variability as well as gradually increase global temperatures. Both of these impacts have the potential to negatively impact the adaptability and resilience of the world's food production capacity; current research indicates climate change is already reducing the productivity of vulnerable cropping systems.

European Federation of Geologists

In 2008, the European Federation of Geologists (EFG) issued the position paper *Carbon Capture and geological Storage*:

The EFG recognizes the work of the IPCC and other organizations, and subscribes to the major findings that climate change is happening, is predominantly caused by anthropogenic emissions of CO_2, and poses a significant threat to human civilization. It is clear that major efforts are necessary to quickly and strongly reduce CO_2 emissions. The EFG strongly advocates renewable and sustainable energy production, including geothermal energy, as well as the need for increasing energy efficiency. CCS [Carbon Capture and geological Storage] should also be regarded as a bridging technology, facilitating the move towards a carbon free economy.

European Geosciences Union

In 2005, the Divisions of Atmospheric and Climate Sciences of the European Geosciences Union (EGU) issued a position statement in support of the joint science academies' statement on global response to climate change. The statement refers to the Intergovernmental Panel on Climate Change (IPCC), as "the main representative of the global scientific community", and asserts that the IPCC represents the state-of-the-art of climate science supported by the major science academies around the world and by the vast majority of science researchers and investigators as documented by the peer-reviewed scientific literature. Additionally, in 2008, the EGU issued a position statement on ocean acidification which states, "Ocean acidification is already occurring today and will continue to intensify, closely tracking atmospheric CO_2 increase. Given the potential threat to marine ecosystems and its ensuing impact on human society and economy, especially as it acts in conjunction with anthropogenic global warming, there is an urgent need for immediate action." The statement then advocates for strategies "to limit future release of CO_2 to the atmosphere and/or enhance removal of excess CO_2 from the atmosphere."

Geological Society of America

In 2006, the Geological Society of America adopted a position statement on global climate change. It amended this position on April 20, 2010 with more explicit comments on need for CO_2 reduction.

Decades of scientific research have shown that climate can change from both natural and anthropogenic causes. The Geological Society of America (GSA) concurs with assessments by the National Academies of Science (2005), the National Research Council (2006), and the Intergovernmental Panel on Climate Change (IPCC, 2007) that global climate has warmed and that human activities (mainly greenhouse gas emissions) account for most of the warming since the middle 1900s. If current trends continue, the projected increase in global temperature by the end of the twentyfirst century will result in large impacts on humans and other species. Addressing the challenges posed by climate change will require a combination of adaptation to the changes that are likely to occur and global reductions of CO_2 emissions from anthropogenic sources.

Geological Society of London

In November 2010, the Geological Society of London issued the position statement *Climate change: evidence from the geological record*:

The last century has seen a rapidly growing global population and much more intensive use of resources, leading to greatly increased emissions of gases, such as carbon dioxide and methane, from the burning of fossil fuels (oil, gas and coal), and from agriculture, cement production and deforestation. Evidence from the geological record is consistent with the physics that shows that adding large amounts of carbon dioxide to the atmosphere warms the world and may lead to: higher sea levels and flooding of low-lying coasts; greatly changed patterns of rainfall; increased acidity of the oceans; and decreased oxygen levels in seawater. There is now widespread concern that the Earth's climate will warm further, not only because of the lingering effects of the added carbon already in the system, but also because of further additions as human population continues to grow. Life on Earth has survived large climate changes in the past, but extinctions and major redistribution of species have been associated with many of them. When the human population was small and nomadic, a rise in sea level of a few metres would have had very little effect on Homo sapiens. With the current and growing global population, much of which is concentrated in coastal cities, such a rise in sea level would have a drastic effect on our complex society, especially if the climate were to change as suddenly as it has at times in the past. Equally, it seems likely that as warming continues some areas may experience less precipitation leading to drought. With both rising seas and increasing drought, pressure for human migration could result on a large scale.

International Union of Geodesy and Geophysics

In July 2007, the International Union of Geodesy and Geophysics (IUGG) adopted a resolution titled "The Urgency of Addressing Climate Change". In it, the IUGG concurs with the "comprehensive and widely accepted and endorsed scientific assessments carried out by the Intergovernmental Panel on Climate Change and regional and national bodies, which have firmly established, on the basis of scientific evidence, that human activities are the primary cause of recent climate change." They state further that the "continuing reliance on combustion of fossil fuels as the world's primary source of energy will lead to much higher atmospheric concentrations of greenhouse gases, which will, in turn, cause significant increases in surface temperature, sea level, ocean acidification, and their related consequences to the environment and society."

National Association of Geoscience Teachers

In July 2009, the National Association of Geoscience Teachers (NAGT) adopted a position statement on climate change in which they assert that "Earth's climate is changing [and] "that present warming trends are largely the result of human activities":

NAGT strongly supports and will work to promote education in the science of climate change, the causes and effects of current global warming, and the immediate need for policies and actions that reduce the emission of greenhouse gases.

Meteorology and Oceanography

American Meteorological Society: The American Meteorological Society (AMS) statement adopted by their council in 2012 concluded: There is unequivocal evidence that Earth's lower atmosphere, ocean, and land surface are warming; sea level is rising; and snow cover, mountain glaciers, and Arctic sea ice are shrinking. The dominant cause of the warming since the 1950s is human activities. This scientific finding is based on a large and persuasive body of research. The observed warming will be irreversible for many years into the future, and even larger temperature increases will occur as greenhouse gases continue to accumulate in the atmosphere. Avoiding this future warming will require a large and rapid reduction in global greenhouse gas emissions. The ongoing warming will increase risks and stresses to human societies, economies, ecosystems, and wildlife through the 21st century and beyond, making it imperative that society respond to a changing climate. To inform decisions on adaptation and mitigation, it is critical that we improve our understanding of the global climate system and our ability to project future climate through continued and improved monitoring and research. This is especially true for smaller (seasonal and regional) scales and weather and climate extremes, and for important hydroclimatic variables such as precipitation and water availability. Technological, economic, and policy choices in the near future will determine the extent of future impacts of climate change. Science-based decisions are seldom made in a context of absolute certainty. National and international policy discussions should include consideration of the best ways to both adapt to and mitigate climate change. Mitigation will reduce the amount of future climate change and the risk of impacts that are potentially large and dangerous. At the same time, some continued climate change is inevitable, and policy responses should include adaptation to climate change. Prudence dictates extreme care in accounting for our relationship with the only planet known to be capable of sustaining human life.

Australian Meteorological and Oceanographic Society

The Australian Meteorological and Oceanographic Society has issued a *Statement on Climate Change*, wherein they conclude: Global climate change and global warming are real and observable ... It is highly likely that those human activities that have increased the concentration of greenhouse gases in the atmosphere have been largely responsible for the observed warming since 1950. The warming associated with increases in greenhouse gases originating from human activity is called the enhanced greenhouse effect. The atmospheric concentration of carbon dioxide has increased by more than 30% since the start of the industrial age and is higher now than at any time in at least the past 650,000 years. This increase is a direct result of burning fossil fuels, broad-scale deforestation and other human activity."

Canadian Foundation for Climate and Atmospheric Sciences In November 2005, the Canadian Foundation for Climate and Atmospheric Sciences (CFCAS) issued a letter to the Prime Minister of Canada stating that We concur with the climate science assessment of the Intergovernmental Panel on Climate Change (IPCC) in 2001 ... We endorse the conclusions of the IPCC assessment that 'There is new and stronger evidence that most of the warming observed over the last 50 years is attributable to human activities'. ... There is increasingly unambiguous evidence of changing climate in Canada and around the world. There will be increasing impacts of climate change on Canada's natural ecosystems and on our socio-economic activities. Advances in climate science since the 2001 IPCC Assessment have provided more evidence supporting the need for action and development of a strategy for adaptation to projected changes

Canadian Meteorological and Oceanographic Society

In November 2009, a letter to the Canadian Parliament by The Canadian Meteorological and Oceanographic Society states: Rigorous international research, including work carried out and supported by the Government of Canada, reveals that greenhouse gases resulting from human activities contribute to the warming of the atmosphere and the oceans and constitute a serious risk to the health and safety of our society, as well as having an impact on all life.

Royal Meteorological Society (UK)

In February 2007, after the release of the IPCC's Fourth Assessment Report, the Royal Meteorological Society issued an endorsement of the report. In addition to referring to the IPCC as "world's best climate scientists", they stated that climate change is happening as "the result of emissions since industrialization and we have already set in motion the next 50 years of global warming – what we do from now on will determine how worse it will get."

World Meteorological Organization

In its *Statement at the Twelfth Session of the Conference of the Parties to the U.N. Framework Convention on Climate Change* presented on November 15, 2006, the World Meteorological Organization (WMO) confirms the need to "prevent dangerous anthropogenic interference with the climate system." The WMO concurs that "scientific assessments have increasingly reaffirmed that human activities are indeed changing the composition of the atmosphere, in particular through the burning of fossil fuels for energy production and transportation." The WMO concurs that "the present atmospheric concentration of CO_2 was never exceeded over the past 420,000 years;" and that the IPCC "assessments provide the most authoritative, up-to-date scientific advice."

Paleoclimatology

American Quaternary Association: The American Quaternary Association (AMQUA) has stated Few credible Scientists now doubt that humans have influenced the documented rise of global temperatures since the Industrial Revolution," citing "the growing body of evidence that warming of the atmosphere, especially over the past 50 years, is directly impacted by human activity.

International Union for Quaternary Research

The statement on climate change issued by the International Union for Quaternary Research (INQUA) reiterates the conclusions of the IPCC, and urges all nations to take prompt action in line with the UNFCCC principles. Human activities are now causing atmospheric concentrations of greenhouse gases — including carbon dioxide, methane, tropospheric ozone, and nitrous oxide — to rise well above pre-industrial levels....Increases in greenhouse gases are causing temperatures to rise...The scientific understanding of climate change is now sufficiently clear to justify nations taking prompt action....Minimizing the amount of this carbon dioxide reaching the atmosphere presents a huge challenge but must be a global priority.

Biology and Life Sciences

Life science organizations have outlined the dangers climate change pose to wildlife.

- American Association of Wildlife Veterinarians
- American Institute of Biological Sciences

In October 2009, the leaders of 18 US scientific societies and organizations sent an open letter to the United States Senate reaffirming the scientific consensus that climate change is occurring and is primarily caused by human activities. The American Institute of Biological Sciences (AIBS) adopted this letter as their official position statement:

The letter goes on to warn of predicted impacts on the United States such as sea level rise and increases in extreme weather events, water scarcity, heat waves, wildfires, and the disturbance of biological systems. It then advocates for a dramatic reduction in emissions of greenhouse gases.

- American Society for Microbiology
- Australian Coral Reef Society
- Institute of Biology (UK)
- Society of American Foresters issued two position statements pertaining to climate change in which they cite the IPCC and the UNFCCC.
- The Wildlife Society (international)

Human Health

A number of health organizations have warned about the numerous negative health effects of global warming

- American Academy of Pediatrics
- American College of Preventive Medicine
- American Medical Association
- American Public Health Association

- Australian Medical Association in 2004 and in 2008
- World Federation of Public Health Associations
- World Health Organization

There is now widespread agreement that the Earth is warming, due to emissions of greenhouse gases caused by human activity. It is also clear that current trends in energy use, development, and population growth will lead to continuing – and more severe – climate change. The changing climate will inevitably affect the basic requirements for maintaining health: clean air and water, sufficient food and adequate shelter. Each year, about 800,000 people die from causes attributable to urban air pollution, 1.8 million from diarrhoea resulting from lack of access to clean water supply, sanitation, and poor hygiene, 3.5 million from malnutrition and approximately 60,000 in natural disasters. A warmer and more variable climate threatens to lead to higher levels of some air pollutants, increase transmission of diseases through unclean water and through contaminated food, to compromise agricultural production in some of the least developed countries, and increase the hazards of extreme weather.

Miscellaneous

A number of other national scientific societies have also endorsed the opinion of the IPCC:

- American Astronomical Society
- American Statistical Association
- Engineers Australia (The Institution of Engineers Australia)
- International Association for Great Lakes Research
- Institute of Professional Engineers New Zealand.

Bibliography

Abbott, Carl. Boosters and Businessmen: *Popular Economic Thought and Urban Growth in the Antebellum Middle West,* Greenwood Press, Westport, CT, 1981.

Anderson, James Richard: *An Economic Geography of Morgan County, Indiana,* Indiana University, 1947.

Battistini, Ren and G. Richard-Vindard: *Biogeography and Ecology in Madagascar,* Junk, The Hague, 1972.

Berry, J.K.: *Beyond Mapping, Concepts, Algorithms and Issues in GIS*, Fort Collins, CO, GIS World Books, 1993.

Carlquist, Sherwin John: *Island Biology*, Columbia University Press, New York, NY, 1974.

Chaklader, Snehamoy: *Sociolinguistics: A Guide to Language Problems in India*, New Delhi, Mittal, 1990.

Clark, Michael J.: *Horizons in Physical Geography*. Basingstoke and London: Macmillan Education, 1987.

Davey, Kenneth J.: *Elements of Urban Management*, Urban Management Program Discussion, The World Bank, Washington DC, 1993.

Dunkerley, Harold B.: *Urban Land Policy, Issues and Opportunities, Introduction and Overview,* Oxford University Press, New York, 1983.

Elangovan K.: *GIS, Fundamentals, Applications and Population Geography,* New India Publishing Agency, New Delhi 2006.

Gressitt, J. Linsley: *Pacific Basin Biogeography. A Symposium. Tenth Pacific Science Congress, Honolulu, Hawaii, 1961,* Bishop Museum Press, Honolulu, HI, 1963.

Haggett, Peter: *Geography: A Modern Synthesis.* New York: Harper and Row, 1975.

Jackson, Peter: *Maps of Meaning: An Introduction to Cultural Geography.* London and Boston: Unwin Hyman, 1989.

James S. Duncan: *Writing Worlds: Discourse, Text, and Metaphor in the Representation of Landscape.* London and New York: Routledge, 1992.

Kunkel, G.: *Biogeography and Ecology in the Canary Islands,* W. Junk, The Hague, 1976.

Lobban, Christopher S.: *Tropical Pacific Island Environments,* University of Guam Press, Mangilao, Guam, 1997.

Lung, Y., and van Tulder, R.: *Cars, Carriers of Regionalism*, Palgrave Macmillan, New York, 2004.

McCarthy, Dennis: *Here be Dragons: How the Study of Animal and Plant Distributions Revolutionized our Views of Life and Earth*, Oxford University Press, Oxford & New York, 2009.

McDonnell, R.A. : *Principles of Geographical Information Systems*, Oxford University Press, Oxford, 1998.

Michael, H.: *City Form and Natural Process: Towards a New Urban Vernacular.* London: Croom Helm, 1984.

Neil, H.: *The Land of Contrasts: 1880-1901,* The American Culture 5. New York: G. Braziller, 1970.

Peter, G.: *The Slow Plague: A Geography of the AIDS Pandemic.* Oxford and Cambridge, MA: Blackwell Publishers, 1993.

Prajapati, R.V.: *A Hand Book of Geography : Realms, Regions and Concepts*, Cyber Tech Pub, Delhi, 2010.

Quaino, Massimo: *Geography and Marxism.* Translated by Alan Braley. Totowa, N.J.: Barnes and Noble, 1982.

Ray Allen: *The Far Western Frontier, 1830-1860.* The New American Nation Series. New York: Harper, 1956.

Robert Fisher and Peter Romanofsky: *Community Organizing for Urban Social Change, A Historical Perspective,* Greenwood Press, Delhi, 1981.

Stephen, D.: *Fields of Vision: Landscape Imagery and National Identity in England and the United States.* Cambridge: Polity, 1993.

Stoddart, D.R.: *Biogeography and Ecology of the Seychelles Islands,* W. Junk, The Hague, 1984.

Tivy, Joy: *Biogeography, A Study of Plants in the Ecosphere,* Oliver and Boyd, Edinburgh, 1971.

Victor, A.: *Geography of Art and Culture*, Contributions to Economic Analysis, Elsevier, 2004.

Watkins, Eric: *The Middle Eastern Environment*, London, The British Society for Middle Eastern Studie, 1995.

Worboys, Michael, and Matt Duckham.: *GIS, a Computing Perspective*, Boca Raton, CRC Press, 2004.

Index

❑❑❑